INTRODUCTION TO ATMOSPHERIC
MODELLING

African Institute of Mathematics Library Series

The African Institute of Mathematical Sciences (AIMS), founded in 2003 in Muizenberg, South Africa, provides a one-year postgraduate course in mathematical sciences for students throughout the continent of Africa. The **AIMS LIBRARY SERIES** is a series of short innovative texts, suitable for self-study, on the mathematical sciences and their applications in the broadest sense.

A complete list of books in the series can be found at www.cambridge.org/mathematics. Recent titles include the following:

AIMS Library Series

INTRODUCTION TO ATMOSPHERIC MODELLING

DOUW G. STEYN

University of British Columbia, Vancouver

CAMBRIDGE
UNIVERSITY PRESS

University Printing House, Cambridge CB2 8BS, United Kingdom

Cambridge University Press is part of the University of Cambridge.

It furthers the University's mission by disseminating knowledge in the pursuit of education, learning and research at the highest international levels of excellence.

www.cambridge.org
Information on this title: www.cambridge.org/9781107499799

First published 2015

Printed in the United States of America by Sheridan Books, Inc.

A catalogue record for this publication is available from the British Library

Library of Congress Cataloging in Publication Data
Steyn, Douw G. (Douw Gerbrand), 1946–
Introduction to atmospheric modelling / Douw G. Steyn, University of British Columbia, Vancouver.
   pages cm. – (African Institute of Mathematics library series)
Includes bibliographical references and index.
ISBN 978-1-107-49979-9
1. Atmospheric models. 2. Atmosphere – Mathematical models. I. Title.
QC861.3.S76   2015
551.501'1–dc23

2014044891

ISBN 978-1-107-49979-9 Paperback

Contents

Prologue

This book has been written specifically for the AIMS Library Series, so its intended audience is students who are attending, have attended, or have backgrounds that would make them eligible to attend the postgraduate programs offered at the African Institute for Mathematical Sciences. The contents of this book could easily be delivered as one of the AIMS postgraduate courses, though it is primarily intended as a self study introductory guide to mathematical modelling in the atmospheric sciences. It has been prepared so that readers with a fairly thorough applied mathematics or physics background can easily, and with little additional reading, understand the main approaches, theoretical and observational underpinnings, intellectual history and challenges of the subject. It is neither a broad introduction to atmospheric science (there exist many such books which serve a very different audience than that intended here), nor is it a review of current research (since that will not serve my intended audience). This book has four distinct, but linked objectives:

- introduce the beauty and wonder of atmospheric phenomena by examining a representative selection;
- explain the importance of scale analysis and scaling arguments in studies of atmospheric phenomena;
- emphasize the power of mathematics in developing an understanding of these phenomena;
- demonstrate how a combination of mathematical modelling, numerical modelling and observations are needed to achieve the understanding.

I start with two rather lengthy introductory chapters designed to introduce the governing equations, their analytical difficulties, and how scale analysis is conducted. The substantive content of this book is organized according to the conventional scale analysis of atmospheric phenomena, and within each scale-specific section I will cover in some detail theoretical (analytical) modelling approaches. Wherever possible and appropriate, I will refer to numerical modelling and observations of the phenomena being discussed. This will be done in order to emphasize the richness of method that characterizes atmospheric science as an academic and professional discipline, but will not constitute a full discussion of atmospheric numerical modelling, or observational meteorology.

Many atmospheric scientists will think that the title implies a book concerned with numerical modelling, and will be surprised that this is not the case. I want to emphasize that intuitive models precede (analytical) mathematical models, which then lead to numerical models. I will not take the second step in that sequence in this book.

In keeping with the spirit of the AIMS Library Series, I will not make extensive reference to research literature, but will rather lean heavily on a small number of selected standard texts listed in my bibliography. These are all texts and colleagues I admire enormously. The colleagues are: Jean-Marie Beckers, Benoit Cushman-Roisin, John Dutton, Solomon Eskinazi and James Holton. I will not include detailed in-text references (since the intended audience will generally not have access to the texts), but will lean heavily on ideas, analyses, approaches and interpretations borrowed from these texts. I here acknowledge the borrowing, and the debt I owe these authors. I acknowledge that any misrepresentations of their ideas are due to my own inadequacies. Furthermore, by this acknowledgement I recognize their ideas as their own, and signal my understanding that not making specific reference leaves me vulnerable to accusations of plagiarism. I am sure they will understand that this has been done because of the nature of books in the AIMS Library Series, and their intended audience. Specifically, Sections 2.1, 4.1, 4.4, 4.5.2 and Chapter 5 follow the approaches taken by Holton, J. R., 1979: *An Introduction to Dynamic Meteorology*, Second Edition, Academic Press, New York. Sections 4.2 and 4.5.1 draw heavily from Cushman-Roisin, B., 1994: *Introduction to Geophysical Fluid Dynamics*, Prentice Hall, NJ. Sections 3.2.1 and 3.2.2 are based on Tennekes,

H., 1973: Similarity laws and scale relations in planetary boundary layers. In D. A. Haugen (Ed.), *Workshop on Micrometeorology*, American Meteorological Society, chapter 4. Section 3.3 uses the approach and results of Haurwitz, B., 1947: Comments on the sea breeze circulation. *Journal of Meteorology*, **40** (1), 1–8. Section 3.2.3 draws on Carson, D. J., 1973: The development of a dry, inversion-capped, convectively unstable boundary layer. *Quarterly Journal of the Royal Meteorological Society*, **99**, 450–467.

I am indebted to a stream of remarkable graduate students I taught at UBC in EOSC 571 (Introduction to Research in Atmospheric Science and Physical Oceanography) over the past eight years. Their enthusiastic and always interesting engagement with the subject material made me think hard about many of the ideas contained in this book. I thank Stefano Galmarini who in a strange way bears ultimate responsibility for this book through first telling me about AIMS, and Fritz Hahne and Barry Green for making possible my stay at AIMS in 2010–2011. Alan Beardon suggested this book, and I thank him for persuading me to take up his idea. David Tranah shepherded the manuscript through the labyrinth of CUP. I have drawn heavily on an excellent summary of dimensional analysis by my colleague George Bluman. Susan Allen read an early version of the manuscript and provided wise and critical advice, and Nadya Moisseeva helped with her excellent work on Sardinian sea breezes. Phil Austin and Nico Fameli changed me from a LaTeX neophyte to LaTeX competent. Ultimately, I take full responsibility for the content and the particular perspective which I bring to the subject matter.

Most of all, I could not have done this without the many years of support and encouragement from Margaret. JoHanna is of course responsible for keeping me humble!

Vancouver, September 2014.

1

Atmospheric phenomena and their study

Earth's atmosphere is a shallow fluid held by gravity to the surface of a spinning sphere whose surface is heated by electromagnetic radiation from the sun. Roughly two thirds of the sphere is covered by water, which continuously undergoes evaporation, condensation, freezing, thawing and sublimation. There is continuous turbulent transport of water vapour and heat between atmosphere and surface. At global scale, the atmosphere is in continuous motion, driven by a relative excess of heating in equatorial regions relative to higher latitudes. The net effect of this motion is a latitudinal redistribution of heat, either directly or by a net transport of moist air from the tropics to higher latitudes where it condenses and falls as precipitation.

Atmospheric large scale motion results in a cascade of energy to smaller scales, producing a complex palimpsest of motion of various types, and at a wide range of scales from global (tens of thousands of kilometres) to a microscale on the order of millimetres. In spatial terms, these motions include some that are quasi two-dimensional, some that are fully three-dimensional, some that are strongly wave-like, and some that are appropriately described as chaotic. Temporally, the motions have time scales of variability that range from astronomically forced variations over tens of thousands of years to turbulent fluctuations of a few seconds in duration, and more recently, decade scale temporal trends driven by human industrial activities. In addition to the purely dynamical phenomena I have just described, the atmosphere includes phenomena whose dynamics are powerfully influenced by thermodynamic processes (such as cloud and precipitation processes) and a wide range of fascinating atmospheric optical phenomena (such as rainbows and circumsolar haloes). This book will primarily concentrate on atmospheric dynamical phenomena, whose time and space scales are graphically shown in Figure 1.1. These phenomena are conventionally grouped into *micro*, *meso* and *macro* scales, and most analyses of atmospheric phenomena focus only on one of the 'scales'. This narrowing of focus has become so sharp that most atmospheric scientists will label themselves according to the 'scale' they study. This three-way scale based classification is more than mere labelling, as it has a profound dynamical basis that will be covered in some detail through this book. At the most basic level, it is in fact a separation of scales, with the implicit assumption that phenomena at one 'scale' operate approximately independently of phenomena at larger and smaller scales. This separation is justified by an analysis of temporal variability in which it is shown that the larger scale phenomena change so slowly that they can be considered as 'frozen' boundary conditions to the phenomena of interest. Similarly, smaller scale phenomena are shown to be 'relaxed' in the sense that they execute many fluctuations during the lifetime of the phenomena of interest, and only their overall (averaged) effect need be considered. In the best of circumstances, it can be shown that a given 'scale' can be treated using an approximation to the full governing equations, and that details of the approximation implicitly or explicitly limit the scales of applicability of the approximate equations.

Space and time scales for atmospheric phenomena included in Figure 1.1 are derived from observations, and the figure is no more than a compact and effective representation of a scale based classification. As will be shown later, there exists an alternative approach to scale definition which employs observed time and space scales and the fundamental equations of atmospheric dynamics. This approach lifts the ideas of scale from a descriptive mode to an analytical mode. There exists yet a third approach to scale analysis that is based on the even more fundamental considerations that lie behind the ideas of dimensional analysis and the *Buckingham pi theorem*, which is presented in the Appendix.

Atmospheric modelling has two fundamental objectives: developing understanding and producing forecasts. The former objective is common to all sciences, and in our specific case involves an analysis of the workings of a model in order to understand the balance of impinging forces or forcings. Obviously, before the force balance analysis is performed, it is essential to establish that the model captures faithfully

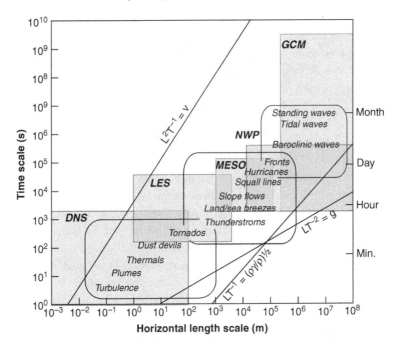

Figure 1.1 Time and length scales of atmospheric dynamical phenomena. The diagonal lines indicate excluded scale ranges which contain phenomena not conventionally treated as part of the atmosphere. In the upper left sector are phenomena in which viscosity dominates, in the lower right sector are phenomena whose characteristic speed is greater than that of sound, as well as phenomena whose acceleration is greater than that due to gravity. The hierarchy of atmospheric numerical models is superposed on the scale classification of phenomena. The models are: Direct Numerical Simulation (DNS), Large Eddy Simulation (LES), Mesoscale (MESO), Numerical Weather Prediction (NWP) and General Circulation Model or Global Climate Model (GCM).

the observed behaviour of the phenomenon. This can only be achieved by comparison of model output with observations of the phenomenon being studied. There is thus no purely theoretical atmospheric modelling, as all modelling is informed by observations, and observations (at least in the last few decades) are based on a theoretical understanding of the phenomena being observed. Weather forecasting is the single most active and important application of atmospheric science, and is no more than a projection forward into time of the space-time variation of atmospheric variables. Because of the importance to society of weather forecasts, quality control consisting of an evaluation of forecast conditions against actually occurring conditions is continuously carried out. This evaluation is again based on observations of the atmosphere. It should thus be evident that atmospheric modelling and observation are intimately linked to each other.

1.1 Models as scientific tools

The term *model* has grown in scientific use in the last few decades. It is used to signify an *abstract analogue* in all sciences, though the levels of abstraction and type of analogue are often specific to particular sciences. It is worthwhile to examine the various uses of the term as it occurs in atmospheric science before setting out to understand atmospheric modelling. In very broad terms, atmospheric models can be Conceptual Models, Analogue Models, Physical Scale Models, Analytical (Mathematical) Models or Numerical Models. These five categories of model are simply different analogues of the actual atmospheric phenomena that are being modelled.

Conceptual models could be called *intuitive models* and are approaches to the understanding of a phenomenon based on the intuition of the scientist. This intuition is developed through education, training and practice of science, and is common to all scientific fields, but the model details will be field specific. The experimental particle physicist will have developed an intuition for the interaction of high energy sub-nuclear particles; an ecologist will have developed an intuitive understanding of the ways organisms or populations interact with each other and their environment; an organic chemist will have developed an intuitive model for chemical reaction mechanisms; an atmospheric scientist will have developed a conceptual model for the evolution of a mid-latitude cyclonic storm, and so on. In many ways, the conceptual models in all fields are developed through the use of the other types of models described here.

Analogue models are employed when one has developed an understanding of a phenomenon in a particular context, and uses that understanding to describe the behaviour of an analogous or similar phenomenon in a different context. The contextual difference may be simply a matter of location or time, or could be rather more profound. An atmospheric example of this is the development of an understanding of the great red spot on Jupiter by analogy with mid-latitude anticyclones in Earth's atmosphere.

Physical scale models are not the primary focus of this book, but a brief explanation of what they involve is needed for completeness. In physical scale models, some or all of the important variables are scaled, either upward or (as is generally the case) downward, in order to overcome difficulties in making observations on the full-scale prototype, or to allow manipulation of a variable that is generally not controllable in the full-scale prototype. There are three major questions in physical scale modelling: technical details needed to construct the scale model; scaling arguments which ensure that the scale model is operating in the same dynamical regime as the full-scale prototype; evaluation of scale model results against the full-scale prototype. Common examples of physical scale models of the atmosphere are wind tunnels (often used to study wind loading on structures, or the dispersion of pollutants around buildings and topographical features) or flow tanks (in which flowing water is used to study atmospheric flow over terrain). A major technical difficulty is incorporating the effects of Earth's rotation in atmospheric scale models. In physical scale models, the scaled version of the prototype is the analogue, and the abstraction is assumptions and approximations that justify the correctness of the scaling.

Analytical (mathematical) models are the primary focus of this book, and their nature, use and meaning will be discussed at some length. In these models, the analogue employed is that measurable physical quantities are represented by variables in a set of mathematical equations, coupled with the assumption that the way the variables in the equation behave, individually and collectively, exactly matches the way the corresponding physical quantities behave. That this is possible can be viewed as either an unfathomable mystery (and therefore akin to an axiom), or simply as the nature of applied mathematics (and therefore no mystery at all). The level of abstraction in an analytical model can generally be identified as the set of assumptions and/or approximations needed to render the complete equations soluble, or at least tractable. This matter is of central importance in analytical modelling of atmospheric phenomena, and will be shown to be closely linked to the definition of scales of phenomena illustrated in Figure 1.1.

Numerical models are numerical implementations of mathematical models. In principle they are no different than analytical mathematical models, and in some cases are simply numerical extensions to analytical models. In practice, numerical models are very different from analytical models because of the nature of the 'solution' they constitute. Rather than the solution being a closed form function, the 'solution' is a large volume of numerical values for all variables over the solution domain. Numerical models are common research and forecasting tools in atmospheric science because the governing equations contain very strong non-linearities that make analytical advances difficult, if not impossible. It should not be a surprise that grid definition, discretization and approximation techniques and control of roundoff errors are enormously complicated and technical questions in atmospheric numerical modelling. In parallel with the classification of atmospheric phenomena into the three scales in Figure 1.1, atmospheric numerical models exist for application at defined scales. Direct Numerical Simulation (DNS) models are used to study fine scale atmospheric motion at the lower left corner of Figure 1.1. Because of computational demands, these models are only employed for study domains up to a few hundred metres in vertical and horizontal extent. Turbulent processes in the lower atmosphere are studied using Large Eddy Simulation (LES) models. Both DNS and LES models are volume averaged implementations of the full equations, and so produce time varying fields. LES models are averaged so as to just resolve the largest energy-carrying turbulent structures (eddies). Mesoscale atmospheric numerical models are based on ensemble averaged equations which have been simplified using the Boussinesq approximation to the full equations. Their output is generally as hourly averages over grid resolutions that span the mesoscale, whose phenomena they are used to study and forecast. All weather forecasts, worldwide, are based on atmospheric numerical models designed to capture atmospheric phenomena at the mesoscale and macroscale. These models are run continuously by national weather forecast agencies, with model output being electronically distributed to regional weather offices for interpretation and the preparation of local forecasts. While the models run continuously, meteorological observations are available on a standard observational cycle. These observations (including satellite derived data) are then ingested into the model runstream in a process known as 4-D data assimilation. The most important examples of large scale atmospheric numerical models are the General Circulation Models (GCMs) used to study global climates over time scales of decades to centuries. Apart from operational weather forecast models, most atmospheric numerical models exist in the research realm and are often available for free download.

This book will focus primarily on analytical models of atmospheric phenomena, but will wherever possible refer to advances that are possible using numerical models. The technical details of atmospheric numerical models will not be covered. There will be no further mention made of atmospheric scale modelling.

The existence of often dominant non-linear processes in the atmosphere means that predictability of atmospheric phenomena must be a vexing question. Over very short times, the dominant processes are part of atmospheric turbulence, and so are essentially unpredictable in detail. Fortunately for weather forecasting professionals, the atmosphere has reasonable predictability over times of three to five days, but very little predictability much beyond that up to seasonal scales when predictability increases. GCMs are able to predict the consequences of various CO_2 emission scenarios with accuracy greater than is possible for the associated human influences and responses to global warming. This varying predictability is simply another aspect of the scale dependence of all atmospheric phenomena.

1.2 Forces in a rotating frame of reference

The atmosphere is bound to, and rotates with, Earth. We observe and analyse the atmosphere relative to this rotating system. It is therefore natural and convenient that the equations of atmospheric motion are stated in this non-inertial frame of reference.

Confining ourselves for the moment to two dimensions, and considering absolute (inertial) and relative (rotating) frames of reference, an absolute velocity vector **U** is given by:

$$\mathbf{U} = \frac{dX}{dt}\mathbf{I} + \frac{dY}{dt}\mathbf{J},$$

where X and Y are coordinates of a point in a fixed frame of reference, and **I** and **J** are unit vectors in that frame. The velocity **U** will have components

$$U = u - \Omega y, \quad V = v + \Omega x$$

in the rotating frame, where Ω is the angular rotation rate of the relative frame of reference[1] and u and v are components of the same velocity

---

[1] $\Omega = 2\pi/86\,400 = 7.272 \times 10^{-5}\ \mathrm{s}^{-1}$.

in the relative frame. These two equations define absolute and relative velocities. Taking a second time derivative of the position vector yields an acceleration vector **A**, which has components:

$$A = a - 2\Omega v - \Omega^2, \quad B = b + 2\Omega u + \Omega^2$$

where a and b are components of the same acceleration in the relative frame. These two equations define absolute and relative accelerations. These equations show that relative acceleration consists of absolute acceleration and two new terms due to the rotating coordinates. The first new term (proportional to Ω) is called the Coriolis acceleration, while the second (proportional to Ω^2) is called the centripetal acceleration. Extension to three dimensions is straightforward since the third axis (that of the rotation) is common to both systems and, in vector form,

$$\mathbf{U} = \mathbf{u} + \mathbf{\Omega} \times \mathbf{r}$$
$$\mathbf{A} = \mathbf{a} + 2\mathbf{\Omega} \times \mathbf{u} + \mathbf{\Omega} \times (\mathbf{\Omega} \times \mathbf{r}),$$

where **r** is the radius vector from Earth's centre, and $\mathbf{\Omega} = \Omega r$ is the Earth rotation rate vector. When the additional two accelerations appear in Newton's second law, they are placed on the right side of the equation and interpreted as virtual forces. As must be clear from the fact that loose matter at Earth's surface does not go flying out into space, gravity overwhelms the centrifugal force (identified with the centripetal acceleration). The centrifugal force does result in Earth having an oblate shape,[2] with the consequence that centrifugal and gravitational forces can be combined into a predominantly gravity component, leaving only the Coriolis force to be accounted for.

Since our analyses are going to be conducted in a local Cartesian framework (in which we replace angular coordinates of longitude/latitude with x/y linear coordinates), we will have to express the Coriolis force in that framework, as depicted in Figure 1.2.

In these coordinates, Earth's rotation vector is:

$$\mathbf{\Omega} = \Omega \cos \phi \mathbf{j} + \Omega \sin \phi \mathbf{k}.$$

The absolute acceleration

$$\frac{d\mathbf{u}}{dt} + 2\mathbf{\Omega} \times \mathbf{u},$$

---

[2] Earth's polar radius is about 6357 km, compared with the equatorial radius of 6378 km.

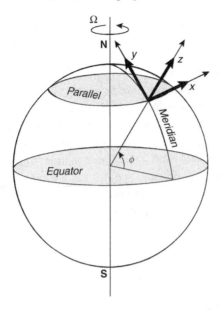

Figure 1.2 A local, Cartesian coordinate system on a spherical Earth: x is East–West, y is North–South, z is vertical and ϕ is the latitude in degrees.

has components:

$$x: \quad \frac{du}{dt} + f_R w - f v$$

$$y: \quad \frac{dv}{dt} + f u$$

$$z: \quad \frac{dw}{dt} - f_R u.$$

We call $f = 2\Omega \sin\phi$ and $f_R = 2\Omega \cos\phi$ the Coriolis and reciprocal Coriolis parameters, respectively. The parameter f is positive in the Northern Hemisphere, zero at the equator and negative in the Southern Hemisphere, whereas f_R is positive everywhere.

1.3 Governing equations

It is worthwhile briefly to consider forces that cause acceleration in fluids since we will be developing an equation for the conservation of momentum of a fluid parcel by a generalization of Newton's second law.

Pressure gradient force

We are all aware that the force we exert on a bicycle pump results in an acceleration of air out of the pump. The force driving this acceleration is due to the gradient of pressure between the body of the pump and the air outside. The pressure gradient force is one of the forces that must be considered when analysing atmospheric motion. This force acts from high pressure to low pressure – down the pressure gradient – and is linear in the gradient itself. The pressure gradient force per unit mass f_{PG} expressed in kinematic terms is thus:

$$\mathbf{f}_{PG} = -\frac{1}{\rho}\nabla p,$$

where p is the pressure field and ρ is the density of air.[3]

Force of gravitation

The force of gravity accelerates air parcels towards the centre of Earth. Atmospheric density decreases geometrically with altitude, and at 100 km above the Earth surface is one millionth of its surface value. Since this height is much smaller than the Earth radius, we can consider the acceleration due to gravity to be constant throughout the atmosphere. The magnitude of this force is given by Newton's law of gravitation, and the force per unit mass $\mathbf{f}_g$ is:

$$\mathbf{f}_g = -g\frac{\mathbf{r}}{r},$$

where $\mathbf{r}$ and r are the Earth radius in vector and magnitude, respectively. As explained in Section 1.2, gravity and centrifugal forces are combined and are represented by a single g. The 'standard' value of g is $g = 9.8066 \text{ ms}^{-2}$, while the actual acceleration due to gravity and centrifugal force is latitude dependent, and varies by about 0.03 ms^{-2}.

Force of friction

Air, like all fluids, has a viscosity, which expresses itself as a force on fluid elements due to distortion of the fluid body. While viscosity is

---

[3] Air is a mixture of nitrogen, oxygen and small amounts of other gases. Its density depends on temperature and pressure, and is $\rho = 1.275 \text{ kgm}^{-3}$ at $0\,^\circ\text{C}$ and 1000 mbar.

ultimately based on interactions between molecules, at a larger scale it can be viewed as quantification of the ability of a fluid to transmit forces laterally due to spatial variations in the mean flow. It is thus the result of relative motion between fluid elements, and can be quantified as a function of the velocity field. In effect, as layers of fluid slide past each other, the molecular attraction between sliding layers results in a drag force between the layers. The force is directed parallel to the layers, and is expressed as a shearing stress (force per unit area). What drives the viscous stresses is thus the gradient of velocity, and it is generally assumed that stresses thus produced are linearly dependent on the velocity gradient. This assumption embodies what is called a Newtonian fluid. For the simple case of one-dimensional (layered) shear flow, the viscous stress in the x direction due to a gradient in the x velocity component in the z direction is:

$$\tau_{zx} = \mu \frac{\partial u}{\partial z}, \tag{1.1}$$

and is expressed in $N\,m^{-2}$ or $kg\,m^{-1}\,s^{-2}$ where μ is the (coefficient of) dynamic viscosity, which is an intrinsic property of the fluid under consideration.[4]

Shearing stress results in a transport of momentum down a velocity gradient. Momentum is passed on from the faster moving layer to the slower. Consider now a cubical parcel of air, as illustrated in Figure 1.3. The parcel is subjected to shearing stress at its top $\tau_{zx,top}$ and bottom $\tau_{zx,bot}$, due to viscous interaction with its neighbouring parcels. The net viscous force per unit mass it experiences is the difference in top and bottom stresses, multiplied by the area over which the stresses act, divided by the mass of the parcel:

$$F_{visc,zx} = \left(\tau_{zx,top} - \tau_{zx,bot}\right) \frac{\Delta x \Delta y}{\rho \Delta x \Delta y \Delta z}.$$

In the limit of an infinitesimally small cubical parcel, $\Delta x, \Delta y, \Delta z \to 0$, $\tau_{zx,top}$ and $\tau_{zx,bot}$ become equal to τ_{zx}, and

$$F_{visc,zx} = \frac{1}{\rho} \frac{\partial \tau_{zx}}{\partial z}.$$

---

4 The viscosity of air is 18.6×10^{-6} Pa s, about 0.02 times that of water.

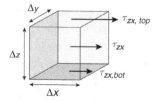

Figure 1.3   Illustration of shearing stress in the x direction on a cubical
volume of air.

From Equation 1.1, and because μ is constant (since it is a property
of the fluid), this becomes:

$$F_{visc,zx} = \frac{1}{\rho}\frac{\partial}{\partial z}\left(\mu\frac{\partial u}{\partial z}\right) = \frac{\mu}{\rho}\frac{\partial^2 u}{\partial z^2}.$$

For convenience, we can define a kinematic viscosity as $v = \mu/\rho$,
which simplifies the expression for $F_{visc,zx}$.

From this formulation, it is evident that if the wind speed is constant with height, or changes linearly with height, the viscous force per
unit mass for vertical shear in the horizontal wind component, $F_{visc,zx}$ is
zero. In general, wind speed will change non-linearly in all three directions, and the viscous force per unit mass will have three components.
In vector notation:

$$\mathbf{F}_{visc} = v\nabla^2\mathbf{u}.$$

In turbulent fluids, the interaction is between between 'parcels' or
eddies of fluid, rather than between molecules. These interactions are
much more effective at transferring momentum than intermolecular
forces. In essence, parcels themselves are translocated in the turbulent
flow, and carry their properties (including momentum) with them. By
analogy, an effective 'eddy viscosity' can be defined. This quantity is
many orders of magnitude larger than v, and is a property of the flow,
rather than the fluid. I will give a brief account of turbulence in fluids in
Sections 2.2.2 and 3.1.

Coriolis force
In the preceding section, we considered the consequence of Earth's rotation, and discovered that doing analysis in a rotating frame of reference

introduces a virtual force called the Coriolis force. This force must also be incorporated into any dynamical analysis of atmospheric motion.

We can now assemble the full set of equations governing atmospheric motion. The first is a set of equations governing fluid motion, and is no more than Newton's second law (simply an expression of the conservation of momentum), set in a rotating framework and expressed for a fluid so that density replaces mass. The equation in notional form is:

Acceleration = Coriolis + Pressure Gradient + Gravity + Friction.

In vector form this is:

$$\frac{\partial \mathbf{u}}{\partial t} + (\mathbf{u} \cdot \nabla)\mathbf{u} = -2\mathbf{\Omega} \times \mathbf{u} - \frac{1}{\rho}\nabla p + \mathbf{g} + \nu\nabla^2\mathbf{u},$$

and in the more directly applicable component form:

$$\frac{du}{dt} = -f_R w + f v - \frac{1}{\rho}\frac{\partial p}{\partial x} + \nu\left(\frac{\partial^2 u}{\partial x^2} + \frac{\partial^2 u}{\partial y^2} + \frac{\partial^2 u}{\partial z^2}\right) \qquad (1.2a)$$

$$\frac{dv}{dt} = -f u - \frac{1}{\rho}\frac{\partial p}{\partial y} + \nu\left(\frac{\partial^2 v}{\partial x^2} + \frac{\partial^2 v}{\partial y^2} + \frac{\partial^2 v}{\partial z^2}\right) \qquad (1.2b)$$

$$\frac{dw}{dt} = f_R u - g - \frac{1}{\rho}\frac{\partial p}{\partial z} + \nu\left(\frac{\partial^2 w}{\partial x^2} + \frac{\partial^2 w}{\partial y^2} + \frac{\partial^2 w}{\partial z^2}\right). \qquad (1.2c)$$

In Equations 1.2a, 1.2b and 1.2c, d/dt is the material derivative since the equations refer to the balance of forces acting on a fluid parcel as it moves with the local flow. Time derivatives are thus made up of two parts: changes due to changing conditions within the fluid parcel being considered, and changes because the fluid parcel is moving through a field that is spatially varying. This defines the material, or advective, derivative:

$$\frac{d}{dt} = \frac{\partial}{\partial t} + u\frac{\partial}{\partial x} + v\frac{\partial}{\partial y} + w\frac{\partial}{\partial z}.$$

In Equations 1.2a and 1.2c, f_R appears multiplied by horizontal and vertical velocities, and as will be shown in Section 2.1, because vertical velocities are very much smaller than horizontal velocities, f_R is neglected in analysis of horizontal motion. Because gravity dominates in equations for vertical motion, f_R is also neglected there, even when multiplied by horizontal velocity, as shown in Section 2.1. From here on we will simply drop f_R from our analysis.

1.4 Boussinesq approximation

Because of the enormously important effect of gravity in the vertical momentum equation, it should by now be clear that processes related to the vertical velocity component deserve particular attention. If we return to Equation 1.2c, before dividing by ρ but dropping the f_R term, and examine the vertical velocity equation, we have:

$$\rho \frac{dw}{dt} = -\rho g - \frac{\partial p}{\partial z} + \mu \left(\frac{\partial^2 w}{\partial x^2} + \frac{\partial^2 w}{\partial y^2} + \frac{\partial^2 w}{\partial z^2} \right).$$

We now decompose vertical velocity w, density ρ and pressure p into reference values and small fluctuations around those values, so that:

$$w = \bar{w} + w', \quad \rho = \rho_0 + \rho', \quad p = \bar{p} + p',$$

which gives

$$(\rho_0 + \rho') \frac{(\bar{w} + w')}{dt} = -(\rho_0 + \rho') g - \frac{\partial (\bar{p} + p')}{\partial z} + \mu \nabla^2 (\bar{w} + w'),$$

where we have written $\nabla^2 (\bar{w} + w')$ for

$$\frac{\partial^2 (\bar{w} + w')}{\partial x^2} + \frac{\partial^2 (\bar{w} + w')}{\partial y^2} + \frac{\partial^2 (\bar{w} + w')}{\partial z^2}.$$

Dividing by ρ_0 and rearranging gives:

$\nu = $ Kinematic viscosity

$$\left(1 + \frac{\rho'}{\rho_0} \right) \frac{d (\bar{w} + w')}{dt} = -\frac{\rho'}{\rho_0} g - \frac{1}{\rho_0} \frac{\partial p'}{\partial z} + \nu \nabla^2 (\bar{w} + w')$$
$$- \frac{1}{\rho_0} \left[\frac{\partial \bar{p}}{\partial z} + \rho_0 g \right].$$

As will be shown in Sections 1.5 and 2.1, the reference state is in hydrostatic equilibrium and so terms in the square bracket must be zero. Additionally, $\rho' \ll \rho_0$, so that $1 + \rho'/\rho_0 \approx 1$, leaving:

$$\frac{d (\bar{w} + w')}{dt} = -\frac{\rho'}{\rho_0} g - \frac{1}{\rho_0} \frac{\partial p'}{\partial z} + \nu \nabla^2 (\bar{w} + w')$$

or

$$\frac{dw}{dt} = -\frac{\rho'}{\rho_0} g - \frac{1}{\rho_0} \frac{\partial p'}{\partial z} + \nu \nabla^2 w.$$

This equation retains the effect of density fluctuations on vertical motion. The procedure whereby we derived this equation is called the

Boussinesq approximation and consists of retaining all terms for density variations in the buoyancy (gravity) term, but neglecting them in the inertia terms. Obviously this only applies to the vertical momentum equation where g appears. In effect, fractional density fluctuations (ρ'/ρ_0) are only retained when multiplied by g. This approximation will be used in all that follows.

Equations 1.2a, 1.2b and 1.2c would completely define the three components of velocity, were it not for the two additional quantities they have introduced, pressure p and density ρ. Fortunately, density and the velocity field are bound together by the principle of conservation of mass, which is expressed as:

$$\frac{\partial \rho}{\partial t} + \frac{\partial}{\partial x}(\rho u) + \frac{\partial}{\partial y}(\rho v) + \frac{\partial}{\partial z}(\rho w) = 0. \tag{1.3}$$

This is called the continuity equation, and expresses the idea that divergence/convergence in a fluid must be balanced by expansion/compression.

Additional constraints are placed on atmospheric motion by the first law of thermodynamics, which expresses the principle of conservation of internal energy. This law requires that the change in internal energy dE must be balanced by heat added or removed dQ and work done dW on or by the parcel of air, so we have

$$dE = dQ + dW. \tag{1.4}$$

The mechanical work done by or on a parcel of air as it expands or contracts is: $dW = -pdV$ where p is the pressure and dV is the change in volume. Changes in internal energy must result in changes in temperature, so that for processes occurring at constant volume, $dE = MC_v dT$ where M is the total mass of air undergoing the change, C_v is the specific heat of air at constant volume and dT is the change in absolute temperature. Since it is convenient to work with changes per unit mass, we divide by M, and Equation 1.4 becomes:

$$dq = C_v dT + pd\alpha, \tag{1.5}$$

where dq is the heat per unit mass, and $d\alpha$ is the change in specific volume (the reciprocal density). Similarly, for processes occurring at constant pressure, we can write:

$$dq = C_p dT + \alpha dp, \tag{1.6}$$

he specific heat of air at constant pressure. Since air
uch like an ideal gas, it must be subject to the equation
pressure, density and temperature:

$$p = \rho \frac{R}{m} T, \tag{1.7}$$

where R is the *gas constant*, and m is the gram molecular weight of air.[5]

Substituting for α in Equations 1.6 and 1.7 leads to a thermodynamic energy equation for the atmosphere:

$$dq = C_p dT - \mathscr{R} T \frac{dp}{p}, \tag{1.8}$$

where $\mathscr{R} = R/m$ is the gas constant per gram molecular weight.

The full set of equations governing atmospheric motion is thus:

$$\frac{du}{dt} = +fv - \frac{1}{\rho}\frac{\partial p}{\partial x} + \nu \left(\frac{\partial^2 u}{\partial x^2} + \frac{\partial^2 u}{\partial y^2} + \frac{\partial^2 u}{\partial z^2} \right) \tag{1.9a}$$

$$\frac{dv}{dt} = -fu - \frac{1}{\rho}\frac{\partial p}{\partial y} + \nu \left(\frac{\partial^2 v}{\partial x^2} + \frac{\partial^2 v}{\partial y^2} + \frac{\partial^2 v}{\partial z^2} \right) \tag{1.9b}$$

$$\frac{dw}{dt} = -\frac{\rho'}{\rho_0}g - \frac{1}{\rho_0}\frac{\partial p'}{\partial z} + \nu \left(\frac{\partial^2 w}{\partial x^2} + \frac{\partial^2 w}{\partial y^2} + \frac{\partial^2 w}{\partial z^2} \right) \tag{1.9c}$$

$$0 = \frac{\partial \rho}{\partial t} + \frac{\partial}{\partial x}(\rho u) + \frac{\partial}{\partial y}(\rho v) + \frac{\partial}{\partial z}(\rho w) \tag{1.9d}$$

$$p = \rho \frac{R}{m} T \tag{1.9e}$$

$$dq = C_p dT - \mathscr{R} T \frac{dp}{p}. \tag{1.9f}$$

As this book is primarily concerned with modelling of atmospheric motion, one might think that the governing equations are sufficient basis for all further analyses. There are however two important topics that have at least indirect bearing on atmospheric dynamics, and so deserve some attention here. These topics are water and electromagnetic radiation, and I will present just the basic ideas in order to provide a reasonable understanding of the bearing water and radiation have on atmospheric dynamics. These topics form rich and interesting fields of

---

[5] $R = 287\,\mathrm{J\,K^{-1}\,kg^{-1}}$. Strictly speaking, air cannot have a molecular weight since it is a mixture. However, it is possible to assign an effective molecular weight $m = 28.97$ to air so that the equation of state works. $C_v = 718\,\mathrm{J\,K^{-1}\,kg^{-1}}$ and $C_p = 1005\,\mathrm{J\,K^{-1}\,kg^{-1}}$.

study in their own right. Most importantly, atmospheric radiation and water are studied using modelling approaches similar in general to those presented here for atmospheric dynamics, but very different in technical detail.

Water in the atmosphere

Water exists in the atmosphere in solid, liquid and gaseous phases. In the gaseous phase, water in the form of water vapour is simply one of the gaseous constituents of the atmosphere. Since the molecular weight of dry air at sea level is 1.6091 times the molecular weight of water vapour, moist air is always less dense than dry air. This density effect clearly has an influence on atmospheric motion because of the centrality of density in the governing equations. Water vapour–air mixtures have a saturation point (dependent on pressure and temperature) at which the maximum possible amount of vapour is present. This determines the maximum density effect of water vapour. The amount of water vapour in air is called the *humidity*, which is quantified by either relative humidity (*RH*) or absolute humidity (*q*):

$$RH = \rho_w/\rho_s,$$

where ρ_w and ρ_s are respectively the density of water vapour and the density of saturated air at the given temperature and pressure. Relative humidity is thus a measure of water vapour content relative to that at saturation. By contrast, absolute humidity is given by:

$$q = m_w/m_a,$$

where m_w and m_a are respectively the mass of water vapour and the mass of dry air in the humid air. Absolute humidity is thus a measure of actual water vapour content of air.

Unsaturated air behaves exactly as dry air, but when air becomes saturated, the possibility exists for water vapour to condense or sublimate to form liquid or solid water suspended in the atmosphere, commonly called clouds. Cloud droplets or crystals form on condensation nuclei and are initially about 0.02 mm in diameter. They can grow by accretion, sublimation or condensation until they are 0.1 mm or larger and fall as precipitation, in its many forms. From a dynamical perspective, phase changes of water in the atmosphere are important because of the enormous amounts of energy absorbed or released through latent heat effects.

Radiation in the atmosphere

All atmospheric motion (like all life on Earth) is ultimately driven by energy derived from the Sun in the form of electromagnetic radiation. The Sun radiates as a black body at a temperature of 5800 K, which results in an average wavelength of about $0.5\,\mu m$. This radiation impinges on the top of the atmosphere with a radiant flux density of $1382\ W\,m^{-2}$. Clearly this energy will be distributed in a way that depends spatially on latitude and temporally on day of the year. Some of this radiation is reflected by clouds, and some is absorbed by the atmosphere, resulting in a reduced amount of radiation reaching Earth's surface. Naturally, the radiation received by a given point on Earth's surface depends on time of day as well as season and latitude. Finally, some radiation falling on the surface will be reflected, and some will be absorbed. This absorption results in a warming of Earth's surface. The effective Earth surface temperature is 290 K which results in an emission of electromagnetic radiation at a mean wavelength of about $10\,\mu m$. Some of the absorbed solar energy is transported into the atmosphere by turbulent motions, some acts to evaporate water which contributes to atmospheric moisture, and some is conducted into Earth. Globally averaged, if these energy exchanges are in balance, the global surface temperature will be constant. Any global imbalance will result in either cooling or warming on a global scale.

There exist strong latitudinal variations in the radiation absorbed and emitted by Earth's surface, with a relative excess in low latitudes. This latitudinal imbalance drives large scale atmospheric motion, and is responsible for much of the weather we experience, especially in mid latitudes. At smaller spatial scales, there exist very strong local differences in radiation absorbed. These local imbalances result in local dynamical phenomena such as land/sea breezes, and slope and valley wind.

1.5 Atmospheric stratification

We have established that the atmosphere is shallow, compared to its horizontal extent. Given this, along with the importance of gravity, the presence of heating/cooling at the surface, and the drag exerted by

Earth's surface on the atmosphere, it should be obvious that many atmospheric properties should be layered. In effect this means that vertical gradients are generally stronger than horizontal gradients. The atmosphere is thus stratified, and a consideration of the effect of stratification on atmospheric motion is warranted. At the scale of the whole atmosphere, a variety of processes are active throughout the atmosphere that result in four distinct layers, as illustrated in Figure 1.4. The lower layer, called the *troposphere,* is characterized by a steady decrease of temperature with height of about 6.5 K km^{-1}. This is because solar radiation absorbed at Earth's surface is distributed vertically by mixing, with some being employed to evaporate water. A combination of this mixing, evaporation and condensation of water, and radiative exchanges result in this rate of temperature decrease. Above the troposphere, temperature increases with height in the *stratosphere*, to result in a temperature maximum at the *stratopause*, at about 50 km elevation. This is due to an absorption of the ultra violet portion of solar radiation by the abundant ozone found in this layer. The final increase in temperature occurs in the *thermosphere* due to absorption of hard ultra violet solar radiation by the very diffuse concentration of oxygen found there. Roughly 50% of atmospheric mass, and nearly all processes we call weather, occur in the troposphere. Below the troposphere lies the highly turbulent *atmospheric boundary layer* whose depth can vary from 2.5 km over heated land surfaces to as low as a few tens of metres over cold oceans or snow and ice surfaces. The atmospheric boundary layer and a corresponding oceanic mixed layer contain almost all life on Earth.

Consider first Equation 1.8 applied to a parcel of air which does not exchange heat with its surroundings. This is called an *adiabatic process,* and can be realized if changes are so fast that conduction and radiation are not important. In such conditions, $dq = 0$ and:

$$C_p dT - \mathscr{R}T\frac{dp}{p} = 0.$$

This can be rewritten as:

$$C_p d \ln T - \mathscr{R} d \ln p = 0.$$

This expression can be integrated from a state defined by pressure p_s and temperature T to a state p and θ and taking logarithms. This gives *Poisson's equation*:

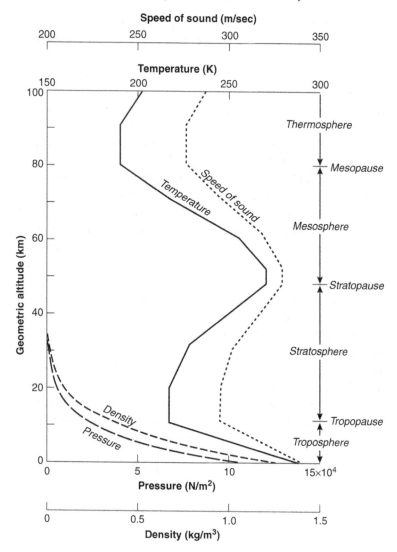

Figure 1.4 Average temperature versus altitude for the large scale atmosphere, showing the four distinct layers.

$$\theta = T \left(\frac{p_s}{p} \right)^{\mathcal{R}/C_p}. \tag{1.10}$$

Here, the *potential temperature* θ is the temperature a parcel of dry air would have if it were compressed or expanded adiabatically to a standard pressure p_s, usually taken as 1000 hPa. Differentiating the logarithm of *Poisson's equation* with respect to height gives:

$$\frac{1}{\theta}\frac{\partial\theta}{\partial z} = \frac{1}{T}\frac{\partial T}{\partial z} - \frac{\mathscr{R}}{C_p p}\frac{\partial p}{\partial z}. \tag{1.11}$$

If gravity balances the vertical component of the pressure gradient force, we have the familiar *hydrostatic balance*:

$$\frac{dp}{dz} = -\rho g.$$

Notice that this is an approximate form for Equation 1.2c for vanishing small vertical velocity, and is an excellent approximation for the vertical distribution of pressure in real atmospheres in all but the most extreme storm conditions. Using the hydrostatic balance and the equation of state, and defining the *lapse rate* of temperature as

$$\Gamma = -\frac{\partial T}{\partial z},$$

Equation 1.11 gives:

$$\frac{1}{\theta}\frac{\partial\theta}{\partial z} = \frac{1}{T}\left(\frac{g}{C_p} - \Gamma\right). \tag{1.12}$$

This means that the potential temperature will be constant in height if $\Gamma = g/C_p \approx 10\,\text{K m}^{-1}$. This is called the *dry adiabatic lapse rate* and is given the symbol Γ_d.

The dry adiabatic lapse rate is a reference lapse rate that defines the static stability of an atmospheric layer with temperature lapse rate Γ. To see this, consider Equation 1.12, which can be rewritten as:

$$\frac{T}{\theta}\frac{\partial\theta}{\partial z} = \Gamma_d - \Gamma.$$

If $\Gamma < \Gamma_d$, potential temperature increases with height, and an air parcel which is moved adiabatically downward/upward from its equilibrium level will experience an upward/downward buoyant force. In this case the forces oppose the motion, and the parcel will tend to return to its equilibrium level. The atmospheric layer is thus termed *stably stratified*. If, however, $\Gamma > \Gamma_d$, potential temperature decreases with height, and an air parcel which is moved adiabatically downward/upward from its equilibrium level will experience a downward/upward buoyant force.

$\Gamma < \Gamma_d$ → Warmer parcel → less dense → downward force

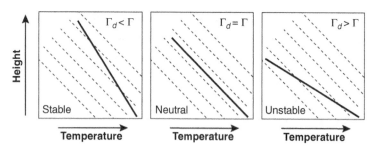

Figure 1.5 Illustration of three possible conditions of temperature lapse
rate (solid line) relative to the dry adiabatic lapse rate (dashed lines).

In this case the buoyancy forces enhance the motion, and the parcel
will tend to be displaced vertically away from its equilibrium level. The
atmospheric layer is thus termed *unstably stratified*. An unstable layer
will be susceptible to vertical overturning due to this buoyancy effect, so
it should be no surprise that at the synoptic scale, the lower atmosphere
is statically stable.

In summary, the static stability criterion is:

$$\frac{dT}{dz} \begin{cases} > \Gamma_d & \text{stable} \\ = \Gamma_d & \text{neutral} \\ < \Gamma_d & \text{unstable.} \end{cases}$$

This is illustrated in Figure 1.5.

As is evident from Figure 1.4, on average the lower atmosphere is
statically stable. This lapse rate was explained as being due to a balance
between solar heating of the surface and evaporation of surface water.
The resultant average lapse rate of 6.5 K km^{-1} is very close to the moist
adiabatic lapse rate.

The case of a stable layer is particularly interesting as stability can
lead to adiabatic oscillations. Consider a parcel vertically displaced by
a small amount δ in a hydrostatic environment. Pressure and density
of the environment ($\bar{p}$ and $\bar{\rho}$) are related by $\bar{\rho}g = -d\bar{p}/dz$. Because of
imbalanced buoyancy forces, the parcel will accelerate in the vertical as:

$$\frac{d^2\delta}{dt^2} = -g - \frac{1}{\rho}\frac{\partial p}{\partial z},$$

where p and ρ are pressure and density of the parcel, respectively.
If parcel pressure adjusts instantaneously to environmental pressure,

$p = \bar{p}$. Using this, and the hydrostatic relationship, the acceleration equation becomes:

$$\frac{d^2\delta}{dt^2} = g\frac{\bar{\rho}-\rho}{\rho} = g\frac{\bar{\theta}-\theta}{\theta}.$$

If the potential temperature at the equilibrium level is θ_0 then the potential temperature after displacement by δ is given by:

$$\theta(\delta) - \theta_0 \approx \left(\frac{\partial\bar{\theta}}{\partial z}\right)_0 \delta,$$

and the acceleration becomes

$$\frac{d^2\delta}{dt^2} = -\left(\frac{g}{\bar{\theta}}\frac{\partial\bar{\theta}}{\partial z}\right)_0 \delta.$$

This represents a vertical oscillation at the frequency N, the Brunt–Väisälä frequency, where:

$$N^2 = \frac{g}{\bar{\theta}}\frac{\partial\bar{\theta}}{\partial z}.$$

In the lower atmosphere, N is in the range 1 to 2×10^{-2} s^{-1}, and the period of the oscillation is roughly 8 minutes. From this, it should be clear that these oscillations, in the presence of a non-zero horizontal wind, will result in wave motion. This oscillation is the origin of a class of waves commonly encountered in the atmosphere, and treated in Chapter 5.

1.6 Atmospheric numerical models

In the early 1900s, L. F. Richardson realized that one strategy for dealing with the complexity and non-linearity of the atmospheric governing equations was to take a numerical rather than analytical approach. At the time, electronic computers were unknown, however with the advent of such computing capability, John von Neumann and his research team pioneered atmospheric numerical modelling in the early 1950s. Since then, the explosive growth in capability and accessibility of computing resources and the development of algorithmic programming languages has led to atmospheric numerical modelling becoming an indispensable and dominant tool, both in forecasting the weather and in providing understanding of atmospheric phenomena at all scales.

While there exist scale-free atmospheric numerical models, it is common practice to use scale specific models. Many examples of fully integrated atmospheric models exist, some developed by large research groups, some developed by scientific consortia, and some developed by the various national weather services. These models form a scale based hierarchy, as illustrated in Figure 1.1.

Atmospheric numerical modelling is a vast and highly technical subject, and is covered in many specialist texts and research papers. Proper treatment of atmospheric numerical models is well beyond the scope and intent of this book. A proper treatment includes consideration of the many possible approximations needed to capture physical effects such as turbulence; the energetics of cloud droplet formation and precipitation; definition of numerical grid structures and transformed spatial coordinates; the filtering of equations to suppress unwanted solutions such as sound waves; the large variety of possible numerical schemes to solve the discretized equations and the appropriate application of boundary and initial conditions. Atmospheric numerical models typically consist of many thousands of lines of computer code, often implemented for parallel execution. Once such a model has been run, one has to deal with many megabytes of output covering all modelled variables over the entire three-dimensional modelling domain. Wherever appropriate, I will point to situations in which the application of an atmospheric numerical model is called for. This will generally be when analytical methods have yielded all they can, but I will not delve into details of models or their application.

2

Scale analysis of the governing equations

Equations 1.9a, 1.9b and 1.9c (more commonly presented without rotational effects) are known as the *Navier–Stokes* equations, and are a notoriously difficult mathematical problem. Demonstrating that they have a smooth, physically reasonable solution is one of the Clay Mathematics Institute's *Millennium Problems*. There are two major difficulties with these equations. In a strange sense, the equations are too complete! They contain as solutions, and cannot discriminate between, the multiplicity of diverse processes that make up the wide spectrum of atmospheric phenomena. Furthermore, there are strong non-linearities embodied in the advective parts (terms of the form $u\partial u/\partial x$) of the material derivative. This non-linearity results in broad-spectrum or multi-scale solutions, and underlies the existence of chaotic behaviour. Our task is to find a way of simplifying the equations through identification of approximations that may render the equations at least partially tractable. As will be seen, the approximate forms will be applicable only to a limited range of scales, which conveniently helps limit the set of phenomena that must be dealt with. In effect the approximations act as a band-pass filter, thereby narrowing our field of view. As will be demonstrated, some of the approximations can be identified with the three scales illustrated in Figure 1.1. While there exist no usefully applicable solutions to the full equations, it is nonetheless possible to make enormous strides in understanding atmospheric phenomena by a careful consideration of scales of phenomena involved, and an analysis of approximate forms of the full equations. In a very real sense, this is an admission that while we do have a complete, unified theory of atmospheric motion, it is not a very helpful theory because of the intractability of the full equations. Fortunately, the equations are enormously helpful, if we consider phenomena according to their scale.

2.1 Order of magnitude analysis

The first approach is to investigate the possibility that some of the terms (hopefully the troublesome ones) in Equations 1.9a, 1.9b and 1.9c can be neglected because of their size relative to the remaining terms. This procedure is conducted by an order of magnitude analysis. As should be clear, it is not possible to conduct a universally applicable analysis of this kind, rather a separate analysis will have to be conducted for specific applications or phenomena. We start by specifying scales for the atmospheric variables. In this context, by a scale, we mean a quantity whose dimensions and numerical value (to within an order of magnitude) match those typically observed in the real atmosphere. We will have to specify values for the magnitudes of variables such as temperature, pressure and density; the magnitudes of fluctuations in these variables; and the length, depth and time scales at which these fluctuations occur. Clearly these quantities are all derived from measurements.

Before specifying the scales, it is important to consider the two broad reasons why there exist dominant scales for atmospheric phenomena.

Forced scales These scales exist because they are forced on the atmosphere by some external variable or variability. The most common examples are the two forced time scales that have their origins in Earth's orbiting around the Sun, and its rotation around its axis. The former introduces an annual time scale into atmospheric phenomena through annually variable solar heating. The latter introduces a diurnal time scale because of diurnally variable heating, and also through the Coriolis force discussed in Section 1.2. Similarly, forced vertical and horizontal space scales arise from the depth of the atmosphere, and Earth's circumference. Smaller forced space scales originate from the length of mountain slopes, the width and depth of valleys and so on.

Free scales These scales are available from measurements, but are not derived from some external forcing, rather they are determined by dynamic processes underlying atmospheric phenomena. For example, the mid-latitude cyclonic storms which are responsible for much of the

Table 2.1 *Horizontal scales and time scales of atmospheric phenomena*

Phenomenon	Horizontal scale (m)	Time scale (s)
Smallest turbulent eddies	10^{-2}–10^{-1}	1
Small turbulent eddies	10^{-1}–1	10
Dust devils	1–10	10^2
Gusts	10–10^2	10^2–10^3
Tornadoes	10^2	10^2–10^3
Cumulonimbus clouds	10^3	10^3–10^4
Land/sea breezes	10^4–10^5	10^3–10^4
Fronts	10^4–10^5	10^5
Cyclonic storms	10^6	10^5–10^6
Planetary waves	10^7	10^6–10^7

world's weather have a horizontal dimension of a few hundred kilometres. While rotation due to Coriolis forces is clearly important here,
so are the typical wind speeds which are in turn driven by pressure
gradients which are expressed over the horizontal extent of the storm.
The circularity of this argument is an indication that the space scale is
internally determined by the dynamical balances that make up the storm
process.

Most important among scales of atmospheric phenomena are the
horizontal space scales and time scales depicted in Figure 1.1. They
are worth repeating in tabular form, see Table 2.1.

The process of scale analysis can best be illustrated by considering
a specific set of phenomena – that of mid-latitude (roughly 35° to 55°
North and South) cyclonic storms. These storms have scales specified in
Table 2.2, and a satellite image of a well-developed mid-latitude cyclone
is shown in Figure 2.1.

In this analysis, we do not distinguish between components of the
horizontal velocity scale. Notice that all the scales are observed, except
the time scale, which is derived from observed scales. As presented
here, the time scale is a time scale for advection, since synoptic scale
phenomena are carried along at the mean wind speed. Scale quantities f
and β will be explained and used in Chapter 4. Similarly, constraints on
the vertical velocity scale can be derived from the governing equations,
as will be shown shortly.

Table 2.2 *Scales of cyclonic storms*

Horizontal velocity scale	U	$10\,\mathrm{m\,s^{-1}}$
Vertical velocity scale	W	$10^{-2}\,\mathrm{m\,s^{-1}}$
Horizontal length scale	L	$10^6\,\mathrm{m}$
Vertical length scale	H	$10^4\,\mathrm{m}$
Time scale	L/U	$10^5\,\mathrm{s}$
Horizontal pressure fluctuation	Δp	$1\,\mathrm{kPa}$
Fractional density fluctuation	$\Delta\rho/\rho_0$	10^{-2}
Reference density	ρ_0	$1\,\mathrm{kg\,m^{-3}}$
Coriolis parameter	f_0	$10^{-4}\,\mathrm{s^{-1}}$
Beta parameter	$df/dy \equiv \beta$	$10^{-11}\,\mathrm{m^{-1}\,s^{-1}}$

Note that for cyclonic storms, the aspect ratio:

$$H/L \ll 1.$$

This feature of the large scale atmosphere was hinted at in Section 1.3, and is true of all larger scale atmospheric dynamical phenomena, which are quasi two-dimensional. In this thin layer, horizontal pressure fluctuations Δp are much smaller than the mean (reference) pressure p_0.[1] As scale decreases, the aspect ratio approaches unity and small scale (primarily turbulent) phenomena change from quasi two-dimensional to fully three-dimensional. This feature raises the expectation that W (the vertical velocity scale) for these phenomena will be comparable to U.

In parallel with pressure fluctuations, density fluctuations are also much smaller than mean density,

$$\rho = \rho_0 + \rho'(x,y,z,t), \quad \rho' \ll \rho_0. \tag{2.1}$$

This observation has important scaling implications that arise out of a consideration of the continuity equation (Equation 1.3), which can be expanded as:

$$\rho_0 \left(\frac{\partial u}{\partial x} + \frac{\partial v}{\partial y} + \frac{\partial w}{\partial z} \right) + \rho' \left(\frac{\partial u}{\partial x} + \frac{\partial v}{\partial y} + \frac{\partial w}{\partial z} \right)$$
$$+ \left(\frac{\partial \rho'}{\partial t} + u\frac{\partial \rho'}{\partial x} + v\frac{\partial \rho'}{\partial y} + w\frac{\partial \rho'}{\partial z} \right) = 0.$$

[1] Pressure is defined as a normal force per unit area, and is thus measured in Newtons per square metre, called Pascals Pa. 1 hectoPascal (hPa) is commonly called a millibar. The atmospheric reference pressure is 1013 hPa, while cyclonic storms have horizontal pressure fluctuations of no more than 20 hPa.

Figure 2.1 A visible band satellite image of a well-developed cyclonic storm off the West Coast of North America on 21 October 2010, showing the characteristic 'comma' cloud band. That rotation is important for this phenomenon is clear from the overall swirling pattern, and curved cloud streaks to the West. The cyclone is about a thousand kilometres in horizontal extent. Vertical extent is not evident in this image. Other scales listed in Table 2.1 must be determined by direct measurement. Image courtesy of NOAA and the University of Washington.

All the terms in the second and third groups (because they contain fluctuating terms) are much smaller than the terms in the first group, and can therefore be dropped from consideration. This reduces the continuity equation to its approximate form:

$$\frac{\partial u}{\partial x} + \frac{\partial v}{\partial y} + \frac{\partial w}{\partial z} = 0.$$

This means that the equation of mass conservation has been reduced to an equation of volume conservation, which is intuitively reasonable if density is approximately constant. This equation also means that sound waves (which are small scale pressure fluctuations) are not important for large scale atmospheric motions. The three terms in this equation are respectively of the following orders of magnitude:

$$\frac{U}{L}, \quad \frac{U}{L}, \quad \frac{W}{H}.$$

If the vertical gradient term were larger than the two horizontal terms, then flow would have to converge in the horizontal. This is impossible because of the rigid bottom on which the atmosphere lies. It is however entirely possible that $W/H \leq U/L$ and because $H \ll L$ we have the condition:

$$W \ll U.$$

The large scale atmosphere is thus quasi two-dimensional, and shallow.

We can now consider the horizontal momentum equations, Equations 1.9a and 1.9b, far from Earth's surface where the effects of friction can be neglected, and only for large scale flow features. The equations and scales of terms are tabulated in Table 2.3.

 As is evident from Table 2.3, terms I to V and VIII are all at least an order of magnitude smaller than terms VI and VII, which are of the same order of magnitude. From this we conclude that, for the phenomena whose scales we are considering, Coriolis and pressure gradient terms must be in approximate balance, resulting in the *geostrophic approximation*:

$$-fv \approx -\frac{1}{\rho}\frac{\partial p}{\partial x}, \quad fu \approx -\frac{1}{\rho}\frac{\partial p}{\partial y}. \tag{2.2}$$

This approximation defines the 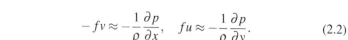 *geostrophic wind* with components:

$$\bar{v}_g \approx \frac{1}{\rho f}\frac{\partial p}{\partial x}, \quad \bar{u}_g \approx -\frac{1}{\rho f}\frac{\partial p}{\partial y}. \tag{2.3}$$

If a spatially varying pressure field is specified, the geostrophic wind field components will also vary in x and y. Equation 2.2 is a diagnostic approximation because it does not allow prognostic (forecast in time) calculations. The geostrophic wind is thus assumed in equilibrium with the driving pressure gradient. For this reason it is often called the *geostrophic balance*. Of course, it is only applicable to large scale

Table 2.3 *Scale analysis of horizontal and vertical momentum equations (all terms in m s^{-2})*

	I	II	III	IV	V	VI	VII	VIII
Horizontal								
x component	$\dfrac{\partial u}{\partial t}$	$+u\dfrac{\partial u}{\partial x}$	$+v\dfrac{\partial u}{\partial y}$	$+w\dfrac{\partial u}{\partial z}$	$+f_R w$	$-fv$	$=-\dfrac{1}{\rho}\dfrac{\partial p}{\partial x}$	$+\nu\dfrac{\partial^2 u}{\partial x^2}$
y component	$\dfrac{\partial v}{\partial t}$	$+u\dfrac{\partial v}{\partial x}$	$+v\dfrac{\partial v}{\partial y}$	$+w\dfrac{\partial v}{\partial z}$		$+fu$	$=-\dfrac{1}{\rho}\dfrac{\partial p}{\partial y}$	$+\nu\dfrac{\partial^2 v}{\partial y^2}$
Scale	$\dfrac{U}{T}$	$\dfrac{U^2}{L}$	$\dfrac{U^2}{L}$	$\dfrac{WU}{H}$	$f_R W$	fU	$\dfrac{\Delta p}{\rho L}$	$\dfrac{\nu U}{L^2}$
Magnitude	10^{-4}	10^{-4}	10^{-4}	10^{-5}	10^{-6}	10^{-3}	10^{-3}	10^{-16}
Vertical								
z component	$\dfrac{\partial w}{\partial t}$	$+u\dfrac{\partial w}{\partial x}$	$+v\dfrac{\partial w}{\partial y}$	$+w\dfrac{\partial w}{\partial z}$	$+f_R u$	$=-\dfrac{1}{\rho}\dfrac{\partial p}{\partial z}$	$-g$	$\nu\dfrac{\partial^2 w}{\partial z^2}$
Scale	$\dfrac{W}{T}$	$\dfrac{UW}{L}$	$\dfrac{UW}{L}$	$\dfrac{W^2}{H}$	$f_R U$	$\dfrac{p_0}{\rho H}$	g	$\dfrac{\nu W}{H^2}$
Magnitude	10^{-7}	10^{-7}	10^{-7}	10^{-8}	10^{-4}	10	10	10^{-15}

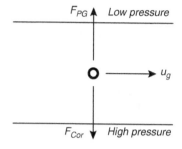

Figure 2.2 Illustration of the geostrophic balance in plan view for a simple pressure gradient. F_{PG} is the pressure gradient force, F_{Cor} is the Coriolis force, and resultant geostrophic wind is u_g.

atmospheric motions which are far enough from Earth's surface that the effects of surface friction are negligible. In practice this means altitudes greater than 1 to 2 km. The geostrophic balance is illustrated in Figure 2.2 for the simplest possible Northern Hemisphere case in which the pressure gradient is constant in magnitude and direction. Notice that the geostrophic wind blows parallel to isobars.

Applying the same scaling approach to the vertical momentum equation results in Table 2.3, which shows that to a high degree of accuracy, the atmosphere is in a state of hydrostatic balance, by which we mean the mean pressure at any point is simply the pressure due to the overlying column of air, we have

$$\frac{1}{\rho_0} \frac{dp_0}{dz} = -g.$$

It is also possible to show that the pressure perturbation field is in hydrostatic equilibrium with the density perturbation field.

2.2 Dimensionless numbers

The Buckingham pi theorem presented in the Appendix shows, in a general way, how dimensionless parameters, denoted π_i in Equation A.3, arise. In atmospheric modelling, there exist a number of such parameters, often called dimensionless governing parameters. These parameters, which are dimensionless numbers, generally arise from a scaling consideration of the equations governing atmospheric flows, and can often be identified as ratios of important terms in the governing equations. Because of their importance, they are named after their discoverers.

2.2.1 The Rossby number

As explained, Equation 2.2 is a diagnostic equation because it assumes an equilibrium, and therefore does not allow forecasts in time. If forces governing the large scale wind field above the influence of surface friction are not in equilibrium, we will have to retain the acceleration terms in Equations 1.9a and 1.9b, resulting in:

$$\frac{du}{dt} - fv = -\frac{1}{\rho}\frac{\partial p}{\partial x}, \quad \frac{dv}{dt} + fv = -\frac{1}{\rho}\frac{\partial p}{\partial y}. \tag{2.4}$$

As demonstrated in Table 2.3, the acceleration terms are an order of magnitude smaller than the geostrophic balance terms. This makes application of these equations problematic, because the acceleration is given by small differences between large terms, leading to severe computational difficulties. Small errors in either of the geostrophic terms will lead to large errors in the calculated acceleration. Nonetheless, this equation leads to an important result. A scaling-level comparison of the magnitude of acceleration compared to the Coriolis force (and therefore pressure gradient force because of the geostrophic balance) may be achieved by forming the ratio of their characteristic scales from Table 2.3

$$\frac{acceleration}{Coriolis} = \frac{U^2/L}{fU} = \frac{U}{fL}.$$

The dimensionless ratio $Ro \equiv U/fL$ is called the Rossby number, and compares the local acceleration to the Coriolis force. The characters of many atmospheric and oceanic flows depend strongly on the value of the Rossby number. This number is the first dimensionless governing parameter we have encountered. It is typical of a handful of such parameters in that it captures, in dimensionless form, an important property of the flow. It is a measure of the importance of rotation relative to acceleration. If Ro is of order unity or smaller, the effects of rotation cannot be neglected. In practical terms, phenomena with a horizontal scale of much less than 100 km are unaffected by Earth's rotation.

2.2.2 The Reynolds number

It would seem important to the horizontal momentum balance to compare the local acceleration (inertial) term to the term due to viscous effects. From Table 2.3 the ratio is:

$$\frac{acceleration}{viscosity} = \frac{U^2/L}{\nu U/L^2} = \frac{UL}{\nu}.$$

The quantity UL/ν is called the *Reynolds number* and given the symbol *Re*. As can be seen from Table 2.3, $Re \approx 10^{12}$, an enormously large number, indicating that to the large scale atmosphere, molecular viscosity is not important, when compared with inertial effects.

Because viscous forces tend to dampen fluctuations in fluid properties, and inertial processes tend to produce such fluctuations through local accelerations, it can be seen that the Reynolds number should inform one about fluid fluctuations. These considerations lead directly into the enormous and difficult topic of turbulence in fluids in general, and in the atmosphere in particular. While fluid flow comprises many flow features, flows can be broadly classified into two classes: *turbulent flows* and *laminar flows*. In laminar flows, viscous effects are effective in damping out fluctuations so that flow patterns are simply organized, streamlines are smooth curves, and analytical mathematical methods yield useful results. By contrast, turbulent flows are characterized by high levels of temporal and spatial (in all three dimensions) fluctuation in all properties, chaotic flow patterns and extremely efficient mixing. This mixing does not occur through molecular diffusion as in laminar flow, but rather is driven by mixing of eddies which carry their properties with them. Turbulence is a property of the flow, rather than of the fluid. Since the Reynolds number is a measure of the relative importance of internal instabilities related to inertial effects (promoting turbulence) and to viscous effects (dampening turbulence), it should come as no surprise that there exists a critical Reynolds number above which flows will be turbulent, and below which they will be laminar. This *critical Reynolds number* lies between 10^3 and 10^4. The atmospheric boundary layer mentioned earlier is in an almost continuous turbulent state.

An important idea in analysis of turbulent flow is the replacement of molecular viscosity and molecular diffusivity by eddy viscosity and eddy diffusivity. These latter quantities are difficult because they are properties of the particular flow being considered, rather than being intrinsic properties of the fluid itself. Because turbulent flows are so efficient at mixing fluid properties (including momentum), eddy viscosity and eddy diffusivity are many orders of magnitude larger than their molecular counterparts. We will return to this topic in Chapter 4 when we discuss turbulence in the lower atmosphere, and the *Ekman layer*.

2.2.3 The Ekman number

Extending the logic of the previous subsection, the remaining pair of forces in Equations 1.9a and 1.9b that should be compared are those due to viscous and Coriolis effects. From Table 2.3 the ratio is:

$$\frac{viscosity}{Coriolis} = \frac{\nu U/L^2}{fU} = \frac{\nu}{fL^2}.$$

The quantity ν/fL^2 is called the *Ekman number* and is given the symbol Ek. As can be seen from Table 2.3, $Ek \approx 10^{-13}$, an extremely small number, indicating that to the large scale atmosphere, molecular viscosity is not important, when compared with rotational effects. If Ro is large, and Ek is small, we have geostrophic motion. However, in conditions in which Ro is very small and Ek is significant, we have an Ekman boundary layer, which will be treated in some detail in Chapter 4.

Since Rossby, Reynolds and Ekman numbers all arise from the horizontal momentum equation, it should not be surprising that there exists a reasonably simple relationship between these three numbers:

$$Ro = \frac{U}{fL} = \frac{UL}{\nu}\frac{\nu}{fL^2} = Re\,Ek.$$

There exist yet more dimensionless numbers that are of importance in studies of the atmosphere, but which are incidental to our present purposes. These include the following.

- The Richardson number, which quantifies the contributions to turbulent kinetic energy from buoyancy relative to wind shear. This number arises from a consideration of equations representing sheared fluids (implying vertical gradients of horizontal wind components) in which density stratification effects are important.
- The Froude number, which is a measure of the importance of stratification relative to inertial effects. This number arises from a

consideration of wave propagation in stratified fluids, and can also be defined in terms of flow speed relative to wave speed.

- The Rayleigh number, which is a measure of the relative importance of heat transfer by convection versus that by conduction.

This chapter has developed a set of approaches to the equations governing atmospheric motion based on the ideas of scales and scaling. This approach was necessary because the equations are mathematically intractable, necessitating a formal way of developing approximations that may reduce the difficulties. The dimensionless numbers resulting form the scaling analysis will be shown to have importance for a wide range of phenomena to be studied in subsequent chapters.

3

Small scale dynamics

As is evident in Figure 1.1, the atmospheric microscale contains turbulent phenomena, and slightly larger coherent phenomena such as plumes, thermals and dust devils. These small scale phenomena are contained in the governing equations, and as was shown for mid-latitude cyclonic storms in Section 2.1, the equations can be simplified by a scale analysis or some other approach that leads to a simpler set of equations, which is only applicable to a restricted set of phenomena at a particular range of scales. We will now examine an important approach that is particularly suited to an analysis of turbulent flows.

3.1 Reynolds decomposition

As has been pointed out, turbulence in fluids is characterized by time dependent, three-dimensional fluctuations of all flow properties. It seems reasonable, therefore, to attempt a separation of the fluctuating part of the flow from the mean flow. In order to do this, some care must be taken with the averaging process that we use to identify the mean flow. Formally, we can only take averages by assuming that it is possible to sample from an infinite set of independent realizations of a specific turbulent flow, all flows having identical initial and boundary conditions, but all being different in detail, because of the inherent statistical nature of turbulence. The set of such flows is called an ensemble, and the average so obtained is called an *ensemble average*. We now decompose the flow into mean and fluctuating parts so that, for velocity components:

$$u = \bar{u} + u', \ v = \bar{v} + v', \ w = \bar{w} + w',$$

37

where $\bar{u}$, $\bar{v}$ and $\bar{w}$ are ensemble averages of the three components. Similarly for pressure, density and potential temperature:

$$p = \bar{p} + p', \; \rho = \bar{\rho} + \rho', \; \theta = \bar{\theta} + \theta'.$$

This appears similar to the approach taken in Equation 2.1, but is fundamentally different because ρ_0 is a reference density, whereas $\bar{\rho}$ is an ensemble average density. Obviously, $\bar{\phi}' = 0$ for all quantities ϕ. We now assume that this procedure, called the *Reynolds decomposition*, allows us to separate the turbulent parts of the flow ϕ' from the mean flow $\bar{\phi}$. The separation is achieved by applying the Reynolds decomposition to, and then taking the time average of, Equations 1.9a, 1.9b and 1.9c. This procedure results in nine new terms, all containing derivatives of averages of products of ϕ' quantities:

$$\begin{bmatrix} \dfrac{\partial \overline{u'^2}}{\partial x} & \dfrac{\partial \overline{u'v'}}{\partial y} & \dfrac{\partial \overline{u'w'}}{\partial z} \\[2ex] \dfrac{\partial \overline{v'u'}}{\partial x} & \dfrac{\partial \overline{v'^2}}{\partial y} & \dfrac{\partial \overline{v'w'}}{\partial z} \\[2ex] \dfrac{\partial \overline{w'u'}}{\partial x} & \dfrac{\partial \overline{w'v'}}{\partial y} & \dfrac{\partial \overline{w'^2}}{\partial z} \end{bmatrix}.$$

The resulting equations, called the *Reynolds equations* are:

$$\frac{d\bar{u}}{dt} = f\bar{v} - \frac{1}{\rho}\frac{\partial \bar{p}}{\partial x}$$
$$+ \left(\nu\frac{\partial^2 \bar{u}}{\partial x^2} - \frac{\partial \overline{u'^2}}{\partial x} \right) + \left(\nu\frac{\partial^2 \bar{u}}{\partial y^2} - \frac{\partial \overline{u'v'}}{\partial y} \right) + \left(\nu\frac{\partial^2 \bar{u}}{\partial z^2} - \frac{\partial \overline{u'w'}}{\partial z} \right)$$

$$\tag{3.1a}$$

$$\frac{d\bar{v}}{dt} = - f\bar{u} - \frac{1}{\rho}\frac{\partial \bar{p}}{\partial y}$$
$$+ \left(\nu\frac{\partial^2 \bar{v}}{\partial x^2} - \frac{\partial \overline{u'v'}}{\partial x} \right) + \left(\nu\frac{\partial^2 \bar{v}}{\partial y^2} - \frac{\partial \overline{v'^2}}{\partial y} \right) + \left(\nu\frac{\partial^2 \bar{v}}{\partial z^2} - \frac{\partial \overline{v'w'}}{\partial z} \right)$$

$$\tag{3.1b}$$

$$\frac{d\bar{w}}{dt} = - \frac{\rho'}{\rho_0}g - \frac{1}{\rho_0}\frac{\partial \bar{p}}{\partial z}$$
$$+ \left(\nu\frac{\partial^2 \bar{w}}{\partial x^2} - \frac{\partial \overline{u'w'}}{\partial x} \right) + \left(\nu\frac{\partial^2 \bar{w}}{\partial y^2} - \frac{\partial \overline{v'w'}}{\partial y} \right) + \left(\nu\frac{\partial^2 \bar{w}}{\partial z^2} - \frac{\partial \overline{w'^2}}{\partial z} \right).$$

$$\tag{3.1c}$$

The quantities $\overline{u'^2}$, $\overline{u'v'}$ ... are called Reynolds stresses because they appear in the Reynolds equations paired with viscous stresses. They are interpreted as terms describing momentum transfer in the fluid due to turbulent eddies carrying their momentum with them as they move through space. Notice that the term $\overline{u'w'}$ is the correlation between u and w velocity fluctuations, and represents u momentum being transported vertically by w fluctuations. Notice also that the Reynolds stresses are symmetric, so $\overline{u'v'} = \overline{v'u'}$, resulting in six, rather than nine, new terms introduced in the equations. If the velocity fluctuations are not correlated, then that particular momentum flux is zero. Measurements of turbulent fluctuations in the atmosphere show that the Reynolds stresses are many orders of magnitude larger than the molecular viscous stresses. Application of the Reynolds decomposition to equations for the conservation of scalars such as moisture or a pollutant result in terms like $\overline{\chi'w'}$, which is the correlation between fluctuations of the scalar quantity χ and vertical velocity w, and represents the scalar being transported vertically by w fluctuations.

As we have noted, vertical gradients are generally larger than horizontal gradients, and this is especially true in that part of the atmosphere immediately overlying Earth's surface, and under statically stable conditions. In these conditions, we can drop all terms except those in $\partial / \partial z$. In addition, $\bar{w}$ and w' will be very small because of proximity to a rigid, near-horizontal lower boundary, and Equations 3.1a, 3.1b and 3.1c reduce to:

$$\frac{d\bar{u}}{dt} = f\bar{v} - \frac{1}{\rho}\frac{\partial \bar{p}}{\partial x} + \left(\nu \frac{\partial^2 \bar{u}}{\partial z^2} - \frac{\partial \overline{u'w'}}{\partial z} \right) \tag{3.2a}$$

$$\frac{d\bar{v}}{dt} = -f\bar{u} - \frac{1}{\rho}\frac{\partial \bar{p}}{\partial y} + \left(\nu \frac{\partial^2 \bar{v}}{\partial z^2} - \frac{\partial \overline{v'w'}}{\partial z} \right) \tag{3.2b}$$

$$\frac{\rho'}{\rho_0}g = -\frac{1}{\rho_0}\frac{\partial \bar{p}}{\partial z}. \tag{3.2c}$$

These simplified equations can be used in many applications in which the effects of turbulence are important, most notably in studies of processes in the lowest layers of the atmosphere. Their application is not without difficulty, since the Reynolds stress terms $\overline{u'w'}$ and $\overline{v'w'}$ have as yet unknown behaviour, and since their introduction in the equations results in more unknowns than equations, they constitute a closure problem.

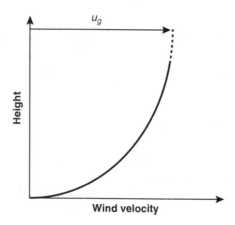

Figure 3.1   Schematic profile of mean wind speed near Earth's surface.

Let us now examine the Reynolds stress $\overline{u'w'}$ in order to understand its physical significance. Consider the very simple case of the atmosphere very near Earth's surface. The mean velocity must go to zero at some very small height, and must tend towards the geostrophic wind speed at some height well above the surface. The profile of wind speed is illustrated in Figure 3.1.

At all heights, the mean horizontal wind speed increases with height so that $\partial \bar{u}/\partial z > 0$. If a parcel of air at height z instantaneously has wind speed $\bar{u}(z)$, and if the instantaneous vertical velocity fluctuation w' is positive (upward), then the parcel of air will be carried upward to height $z + \delta$. This parcel of air has a horizontal momentum of $\rho \bar{u}(z)$, but arrives at a level where the mean horizontal momentum is $\rho \bar{u}(z + \delta)$, and so is in a state of momentum deficit relative to the surrounding air. This will decelerate air at level $z + \delta$ by an amount $-u'$. This retardation is caused by the Reynolds stress, which is equal to the mean momentum deficit, $-\rho \overline{u'w'}$, and is in the same direction as the (molecular) viscous stress $\nu \partial \bar{u}/\partial z$. The same argument holds for $\bar{v}$. In summary:

$$\frac{\partial \bar{u}}{\partial z} > 0, \quad +w' \Rightarrow -u' \quad \text{and} \quad -w' \Rightarrow +u', \ \overline{u'w'} < 0$$

$$\frac{\partial \bar{u}}{\partial z} < 0, \quad +w' \Rightarrow +u' \quad \text{and} \quad -w' \Rightarrow -u', \ \overline{u'w'} > 0.$$

This relationship between $\partial \bar{u}/\partial z$ and $\overline{u'w'}$ leads naturally to:

$$\overline{u'w'} = -v_e \frac{\partial \bar{u}}{\partial z} \quad \text{and} \quad \overline{v'w'} = -v_e \frac{\partial \bar{v}}{\partial z} \qquad (3.3)$$

where v_e is an *eddy viscosity*, and quantifies momentum transport due to turbulent eddies. This approach is called *first order closure* since it replaces a second order statistic $\overline{u'w'}$ with a first order term $\partial \bar{u}/\partial z$. As pointed out earlier, v_e is many orders of magnitude larger than v, and is a property of the flow, rather than the fluid. Of course the problem has not fully disappeared as the functional dependence of v_e on height and possibly other atmospheric properties has yet to be determined.

With this in mind, it is worthwhile to return to Equations 3.2a and 3.2b, and consider the bracketed terms in the u and v equations:

$$\left(v \frac{\partial^2 \bar{u}}{\partial z^2} - \frac{\partial \overline{u'w'}}{\partial z} \right) \quad \text{and} \quad \left(v \frac{\partial^2 \bar{v}}{\partial z^2} - \frac{\partial \overline{v'w'}}{\partial z} \right)$$

can be rewritten as

$$\frac{\partial}{\partial z} \left(v \frac{\partial \bar{u}}{\partial z} - \overline{u'w'} \right) \quad \text{and} \quad \frac{\partial}{\partial z} \left(v \frac{\partial \bar{v}}{\partial z} - \overline{v'w'} \right).$$

Implementing the first order closure gives:

$$\frac{\partial}{\partial z} \left(v \frac{\partial \bar{u}}{\partial z} + v_e \frac{\partial \bar{u}}{\partial z} \right) \quad \text{and} \quad \frac{\partial}{\partial z} \left(v \frac{\partial \bar{v}}{\partial z} + v_e \frac{\partial \bar{v}}{\partial z} \right),$$

and because $v \ll v_e$, Equations 3.2a, 3.2b and 3.2c become

$$\frac{d\bar{u}}{dt} = f\bar{v} - \frac{1}{\rho} \frac{\partial \bar{p}}{\partial x} + \frac{\partial}{\partial z} \left(v_e \frac{\partial \bar{u}}{\partial z} \right) \qquad (3.4a)$$

$$\frac{d\bar{v}}{dt} = -f\bar{u} - \frac{1}{\rho} \frac{\partial \bar{p}}{\partial y} + \frac{\partial}{\partial z} \left(v_e \frac{\partial \bar{v}}{\partial z} \right) \qquad (3.4b)$$

$$\frac{\rho'}{\rho_0} g = -\frac{1}{\rho_0} \frac{\partial \bar{p}}{\partial z}. \qquad (3.4c)$$

Notice that v_e must remain inside the derivative since we do not (yet) know its dependence on z. These equations can be used as a substitute for Equations 3.2a, 3.2b and 3.2c in circumstances where v_e is known. In practice, v_e is specified as a function of height and atmospheric stability, based on measurements of mean velocity gradients and momentum fluxes. This process of replacing an unknown variable with approximate forms derived from measurements is known as *parameterization*.

As will be noticed, we have not yet managed to develop an approximation at either large or small scale that has resulted in an analytically tractable form for the governing equations. We have however developed methods that have considerably simplified the full equations, and have at the same time developed some insight into atmospheric dynamics at both large and small scales. In the coming sections we will see the kind of simplification and idealization that is needed to develop analytically tractable sets of equations that give reasonably realistic representations of specific atmospheric phenomena.

3.2 The atmospheric boundary layer

The *atmospheric boundary layer* is the atmospheric layer closest to Earth's surface, and is in an almost continuously turbulent state. This layer is strongly affected by fluxes of momentum, heat and moisture. The layer has variable thickness, in both space and time, depending on the magnitude of these fluxes, and can have different character, depending on the dominance of either heat flux or momentum flux. In this section we will examine the height dependence of the atmospheric boundary layer under two strongly limiting sets of conditions. While these conditions are so limited as to be almost unrealistic, it will be seen that the results of our analysis are very revealing of what happens in realistic conditions.

3.2.1 The logarithmic wind profile

Of considerable interest is the vertical dependence of the mean wind (schematically shown in Figure 3.1), since that is what mixes pollutants, exerts stress on buildings, plants and other organisms, and in many ways conditions the environment. Under all but the most extreme conditions, the acceleration terms in Equation 3.2a are negligible when compared to the magnitudes of the remaining terms, so we can safely assume stationary conditions. If wind speeds are so strong that heating effects due to radiation or latent heat exchanges can be neglected, then the atmosphere will be statically neutral (adiabatic). In practice this means that the mean wind speed measured well above the surface must be greater than 10 m s^{-1}. Furthermore, if we assume there are no significant horizontal variations in surface or atmospheric conditions, the atmosphere becomes horizontally homogeneous, and the problem collapses to one dimension (z). An additional condition is also needed here. We assume the atmosphere is *barotropic* which means that density is a function of pressure alone $\rho \equiv \rho(p)$, which makes the geostrophic wind independent of height. This condition will be explained in more detail in Chapter 4.

Under the stationarity and homogeneity assumptions, Equations 3.2 thus become:

$$f\bar{v} = \frac{1}{\rho}\frac{\partial \bar{p}}{\partial x} - \left(v\frac{\partial^2 \bar{u}}{\partial z^2} - \frac{\partial \overline{u'w'}}{\partial z}\right)$$

$$f\bar{u} = \frac{1}{\rho}\frac{\partial \bar{p}}{\partial y} + \left(v\frac{\partial^2 \bar{v}}{\partial z^2} - \frac{\partial \overline{v'w'}}{\partial z}\right) \qquad (3.5)$$

$$g = -\frac{1}{\rho}\frac{\partial \bar{p}}{\partial z}.$$

The mean wind speed near the surface will be zero. The height at which this occurs is called the *roughness length*, z_0. Well away from the surface the wind will be the geostrophic wind speed U_g, with components $\bar{u}_g$ and $\bar{v}_g$ given by Equation 2.3. These can be substituted into Equation 3.5 to give:

$$f(\bar{v} - \bar{v}_g) = -\frac{d}{dz}\left(v\frac{d\bar{u}}{dz} - \overline{u'w'}\right)$$

$$f(\bar{u} - \bar{u}_g) = \frac{d}{dz}\left(v\frac{d\bar{v}}{dz} - \overline{v'w'}\right),$$

where $\bar{u}$ and $\bar{v}$ are height dependent horizontal wind components in the atmospheric boundary layer. Furthermore, as already discussed, molecular viscosity stresses are negligible compared to eddy viscous stresses, and can be dropped to yield the so-called *velocity deficit law*:

$$f(\bar{v} - \bar{v}_g) = -\frac{d}{dz}\left(-\overline{u'w'}\right)$$

$$f(\bar{u} - \bar{u}_g) = \frac{d}{dz}\left(-\overline{v'w'}\right). \qquad (3.6)$$

It is wise to retain the negative sign inside the derivative since both velocity correlation terms are negative. Note that we have not committed ourselves to the first order closure yet, and in fact the eddy stress terms on the right hand side of Equation 3.6 are the major difficulty. The kinematic surface eddy shearing stress ($\tau_x/\rho = |\overline{u'w'}|$) can be used to define a velocity scale called the *surface friction velocity*, u_*, by:

$$\frac{\tau_x}{\rho} = u_*^2 = \sqrt{\overline{u'w'}^2 + \overline{v'w'}^2}.$$

The upper boundary conditions for Equation 3.6 are

$$\bar{u} \to \bar{u}_g, \quad \bar{v} \to \bar{v}_g$$
$$-\overline{u'w'} \to 0, \quad -\overline{v'w'} \to 0,$$

and the lower boundary conditions are

$$\bar{u}(z_0) = 0, \quad \bar{v}(z_0) = 0$$
$$-\overline{u'w'}(z_0) \to u_*^2, \quad -\overline{v'w'}(z_0) \to 0.$$

Notice that we impose a no-slip condition at z_0, the lower boundary, even though viscosity (which is the usual source of this condition) does not appear in the equations.

Let us now focus on the lowest layers of the atmospheric boundary layer. Here, the (no-slip) velocity boundary condition requires Equation 3.6 to collapse to:

$$f\bar{v}_g = \frac{d}{dz}\left(-\overline{u'w'}\right), \quad -f\bar{u}_g = \frac{d}{dz}\left(-\overline{v'w'}\right). \tag{3.7}$$

Which means that the vertical gradients of the Reynolds stresses are at their maximum at z_0. For the lower parts of this layer (called the surface layer), the length scale is z_0 and the velocity scale is u_*. Using these scales to non-dimensionalize Equation 3.7 gives:

$$\frac{z_0 f\bar{v}_g}{u_*^2} = \frac{d}{d\zeta_0}\left(-\frac{\overline{u'w'}}{u_*^2}\right), \quad -\frac{z_0 f\bar{u}_g}{u_*^2} = \frac{d}{d\zeta_0}\left(-\frac{\overline{v'w'}}{u_*^2}\right), \tag{3.8}$$

where $\zeta_0 = z/z_0$.

Using the values $z_0 = 10^{-2}$ m, $U_g = 10$ m s^{-1}, $u_*^2 = 10^{-1}$ m^2 s^{-2} and $f = 10^{-4}$ s^{-1}, gives an upper bound of 10^{-4} to the left hand sides of Equation 3.8, a value so small that for the surface layer, where $\zeta_0 = z/z_0$ has finite values, the vertical stress gradients are effectively zero. This means that the surface layer is a layer of constant stress. With an appropriate rotation of the coordinate system, we have:

$$-\overline{u'w'} = u_*^2, \quad -\overline{u'w'} = 0.$$

Because wind speeds in this layer are so low, and length scales so small, Coriolis effects can be neglected, and the wind profile $\bar{u}(z)$ must scale by only z_0 and u_*, and can be written as:

$$\frac{\bar{u}}{u_*} = f_x(\zeta_0), \quad \frac{\bar{v}}{u_*} = 0. \tag{3.9}$$

This is the so-called *law of the wall*. Clearly this scaling cannot apply in the upper parts of the atmospheric boundary layer. We now return to Equation 3.6 and seek a scaling applicable to the upper boundary. Here, z_0 cannot be the length scale, rather the depth of the entire layer must be important. Call this depth h. Velocities must still be scaled by u_* since it is the surface (eddy) drag that drives velocity gradients. Using these scales, Equation 3.6 becomes:

$$-\frac{hf}{u_*}\left(\frac{\bar{v}-\bar{v}_g}{u_*}\right) = h\frac{d}{dz}\left(\frac{-\overline{u'w'}}{u_*}\right)$$
$$\frac{hf}{u_*}\left(\frac{\bar{u}-\bar{u}_g}{u_*}\right) = h\frac{d}{dz}\left(\frac{-\overline{v'w'}}{u_*}\right). \tag{3.10}$$

We are free to choose h in any convenient way. The value $h = \eta u_*/f$ seems obvious as it is consistent with this scaling, and observations indicate it works well with values of η in the range 0.3 to 0.4. The result is that there are no free parameters in the problem, as expressed at the upper boundary conditions. Notice that these equations do not hold in the surface layer, where the law of the wall must apply. This means that Equation 3.10 can be written:

$$\frac{\bar{v}-\bar{v}_g}{u_*} = F_y(\zeta_h), \quad \frac{\bar{u}-\bar{u}_g}{u_*} = F_x(\zeta_h) \tag{3.11}$$

where $\zeta_h = zf/u_*$.

While it is clear that Equations 3.9 and 3.11 are valid for different parts of the atmospheric boundary layer, it should also be clear that there should be some intermediate range of heights within the boundary layer where they must hold simultaneously, in the sense that their solutions must match. This matching is achieved by requiring the wind shear in the upper part of the layer to match that in the lower part. We therefore take the derivatives of Equations 3.9 and 3.11 with respect to their dimensionless arguments, $\zeta_0 = z/z_0$ and $\zeta_h = zf/u_*$, and for convenience, multiplying by z/u_*:

$$\frac{z}{u_*}\frac{d\bar{u}}{dz} = \zeta_0\frac{df_x}{\zeta_0} = f_m(\zeta_0) \tag{3.12}$$

$$\frac{z}{u_*}\frac{d\bar{u}}{dz} = \zeta_h\frac{dF_x}{\zeta_h} = F_m(\zeta_h). \tag{3.13}$$

In this matching layer, $(z/u_*)\,d\bar{u}/dz$ must simultaneously be functions of both ζ_0 and ζ_h so that Equations 3.12 and 3.13 must match under the double limit $\zeta_0 \to \infty$ and $\zeta_h \to 0$. Apart from the trivial or meaningless possibilities that $f_m, F_m \to 0$ or ∞, the only possibility is that both matching functions become independent of their arguments as the double asymptotes are approached. This means that they must approach the same universal constant. So, for $\zeta_0 \gg 1$ and $\zeta_h \ll 1$ the dimensionless wind shear must be:

$$\frac{z}{u_*}\frac{d\bar{u}}{dz} = \frac{1}{\kappa}, \tag{3.14}$$

where κ is the *von Karman constant*, whose value can only be determined by measurement of $\bar{u}(z)$. Many measurement campaigns concentrating on the atmospheric surface layer over carefully selected sites result in a value of $\kappa \approx 0.4$. This formulation is in agreement with measurements again from many locations, over many different surface types, which show that the mean wind vertical gradient declines linearly in height.

It is worth pointing out that this equation implies that the eddy viscosity has linear height dependence in the surface layer, and is given by:

$$v_e = \kappa u_* z.$$

Integrating Equation 3.14 with respect to z gives the *logarithmic wind profile*

$$\bar{u}(z) = \frac{u_*}{\kappa z}\ln\left(\frac{z + z_0}{z_0}\right),$$

where z_0, the roughness length, can now be seen as a constant of integration chosen so that the lower boundary condition is exactly met: $\bar{u} = 0$ at $z = z_0$. The roughness length depends on the physical roughness of the surface and has values shown in Table 3.1.

This analysis, which started off from the governing horizontal momentum equations under highly restrictive conditions, has produced a very important, practical result – the logarithmic wind profile. Further analysis is needed to develop an understanding of the mean horizontal wind speed profile under diabatic conditions. This analysis requires a rather more detailed treatment of boundary layer turbulence than is appropriate for this book. Most interesting is that the diabatic profiles are simply modified forms of the adiabatic (logarithmic) profile. Details

Table 3.1 *Roughness length for various surface types, x/H is the aspect ratio of roughness elements*

Surface type	z_0 (m)
Open sea; fetch of at least 5 km	0.0002
Mud flats, snow; no vegetation, no obstacles	0.005
Open flat terrain; grass, few isolated obstacles	0.03
Low crops; occasional large obstacles, $x/H > 20$	0.10
High crops; scattered obstacles, $15 < x/H < 20$	0.25
Parkland, bushes; numerous obstacles, $x/H \approx 10$	0.5
Irregular large obstacles (suburb, forest)	1.0
City centre with high- and low-rise buildings	≤ 2.0

of the mean horizontal wind profile above the surface layer will be examined in the following section.

3.2.2 The Ekman layer

Lying immediately above the surface layer, is a layer which is still fully turbulent, but is governed by the Coriolis effect, the geostrophic wind aloft, and friction from below. We have already introduced the equations for this layer – the velocity deficit law represented by Equation 3.6. We retain the simplifying assumptions of stationarity, horizontal homogeneity, adiabatic conditions and a barotropic atmosphere. A natural simplification is to align the coordinate system so that the geostrophic wind has only one component. We set $\bar{v}_g = 0$, and have:

$$f\bar{v} = -\frac{d}{dz}\left(-\overline{u'w'}\right)$$
$$f(\bar{u} - \bar{u}_g) = \frac{d}{dz}\left(-\overline{v'w'}\right). \tag{3.15}$$

As before, pressure is a function of horizontal position only, allowing us to define a geostrophic wind speed. The troublesome Reynolds stresses remain in both equations. If we commit ourselves to a first order closure by replacing both $\left(-\overline{u'w'}\right)$ and $\left(-\overline{v'w'}\right)$ with ν_e times the respective vertical velocity gradient as before, we have:

$$f\bar{v} = -v_e \frac{d^2\bar{u}}{dz^2}$$

$$f(\bar{u} - \bar{u}_g) = v_e \frac{d^2\bar{v}}{dz^2}.$$

(3.16)

Notice that this formulation implies that v_e is not dependent on height. While this is not true in the surface layer, it might be helpful in this layer, so we press on. The boundary conditions we apply here are:

$$z = 0: \quad \bar{u} = 0, \quad \bar{v} = 0$$

$$z \to \infty: \quad \bar{u} = \bar{u}_g, \quad \bar{v} = 0.$$

Note that the first of these conditions appears to be technically in conflict with the lower boundary condition that defined the surface layer, since that had horizontal wind speed going to zero at $z = z_0$, not $z = 0$. In practice this is not a difficulty since in the layer we are considering, $z/z_0 \gg 1$.

We now seek solutions to Equation 3.16 of the form $u = c_1 + c_2 \exp(c_3 z)$. Solutions to the characteristic equation are

$$c_3 = \pm(1 \pm i)\frac{1}{d},$$

where the length d is

$$d = \sqrt{\frac{2v_e}{f}}.$$

If, for simplicity, we restrict our solutions to the Northern Hemisphere where f is positive, and apply the top boundary condition which excludes exponentially growing solutions, and the lower boundary condition, we have:

$$\bar{u} = \bar{u}_g [1 - \exp(-z/d)\cos(z/d)]$$

$$\bar{v} = \bar{u}_g \exp(-z/d)\sin(z/d).$$

(3.17)

In Figure 3.2, we plot the two wind components $\bar{u}$ against $\bar{v}$ for different heights illustrating what is called the *Ekman spiral*, after its discoverer V. W. Ekman. This solution is notable for a number of reasons.

Firstly, the distance d, called the *Ekman depth*, can be used to give an estimate of the depth of the neutral atmospheric boundary layer. Notice from Figure 3.2 that velocity components change only very slightly for

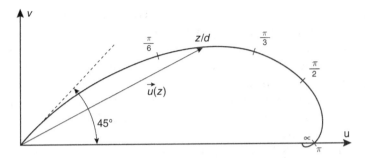

Figure 3.2 Wind velocity components defining the Ekman spiral in the Northern Hemisphere. Values of z/d are given along the spiral.

$z/d \geq \pi$. It seems reasonable to assume that the layer depth can be estimated by $z \approx \pi d$. It is observed that the depth of the statically neutral boundary layer in strong wind conditions is roughly 1 km. A given value of $f \approx 10^{-4}$ s^{-1}, leads to a value of $v_e \approx 5$ m^2 s^{-1} for the eddy viscosity.

This solution also illustrates the apparent oddity that at some height in the boundary layer, there is a flow component perpendicular to the driving geostrophic wind. As we approach the bottom, the flow is at $45°$ to the geostrophic wind. Also counter intuitive is the discovery that at some height near d, the mean wind is actually greater than the geostrophic wind.

We earlier defined the Ekman number as

$$Ek = \frac{v}{fL^2},$$

which leads to:

$$\frac{d}{L} \approx \sqrt{Ek},$$

if L is a horizontal length scale of the flow, and momentum transport is by Reynolds stresses, rather than molecular viscosity. The Ekman number can therefore be interpreted as the square of the atmospheric boundary layer depth to the horizontal scale of the flow driving boundary layer mixing.

We have already seen that the eddy viscosity increases linearly in the surface layer from a value of 0 at $z = z_0$. It seems unreasonable therefore that v_e is constant through the entire Ekman layer. This is in fact true, and observations indicate that v_e does increase linearly through the lower boundary layer to a local maximum at approximately one third of the boundary layer depth, and thereafter decreases slowly to near zero where turbulence ceases at the boundary layer top. That details of the Ekman layer formulation are a great simplification should not be a surprise, since the equations are valid only under highly restrictive conditions. This does not detract from the utility of the Ekman analysis. It has provided considerable insight into mechanics of the adiabatic atmospheric boundary layer. The Ekman wind spiral has been observed in many locations, though not as simply as depicted in Figure 3.2. Most interestingly, Ekman's analysis of the Ekman spiral in the upper layers of the ocean provided an explanation of the arctic explorer Fridtjof Nansen's observation that icebergs drift at an angle of 20° to 40° to the right of the mean wind direction in the Northern Hemisphere.

3.2.3 The convective boundary layer depth

The preceding section dealt with atmospheric boundary layers in which mixing is dominated by turbulence generated by wind shear, and all processes are adiabatic. Under conditions in which mixing is dominated by turbulence generated by heating of the underlying Earth surface, a completely different approach is needed. Strong solar heating at Earth's surface occurs under clear sky conditions. These commonly occur when the large scale atmosphere is dominated by a high pressure system (called an *anticyclone*). Anticyclonic weather is characterized by a slow sinking (subsidence) of the large scale atmosphere. This sinking results in adiabatic heating of the lower troposphere, and hence the presence of a deep temperature inversion, forming a layer of statically stable air overlying the atmospheric boundary layer. In addition, the subsidence induced warming acts to suppress clouds by evaporation of cloud liquid water, thus increasing solar heating of the surface. At the surface, however, solar heating results in a decline of potential temperature with height, and a shallow layer of strongly unstable air immediately above the surface. In the surface layer, upward fluxes of sensible heat drive strong turbulent mixing that penetrates through the entire boundary layer and interacts strongly with the capping inversion aloft. This interaction causes downward mixing of inversion air, and a downward flux of sensible heat at the inversion base. Idealized profiles of potential temperature and turbulent heat flux are shown in Figure 3.4. As can be seen from these profiles, mixing in the convective boundary layer is so strong that gradients of potential temperature vanish. Indeed, wind speed, moisture and pollutants all have no gradients in this layer. For this reason it is often called the *mixed layer*. The inversion-capped mixed layer is enormously important from a pollution perspective as it defines the layer into which pollutants are mixed and constrained. Far from individual source effects, the mean concentration of pollutants is determined by the strength of the pollution source, mean wind and the mixed layer depth. Figure 3.3 illustrates a well-developed convective boundary layer. In Section 3.1 we saw how terms such as $\rho \overline{u'w'}$ represented a transport of u momentum in the w direction because of turbulent fluctuations mixing air in a layer that had vertical gradients of u. In an analogous way, $\rho C_p \overline{\theta'w'}$ represents a transport of sensible heat by turbulent fluctuations in a layer with vertical gradients of potential temperature. It is conventional to work with the kinematic turbulent sensible heat flux $\overline{\theta'w'}$.[1]

The enthalpy equation for the atmospheric boundary layer, in differential form equates the rate of change of temperature to the vertical convergence of turbulent sensible heat flux density:

$$\rho C_p \left(\frac{d\bar{\theta}}{dt} \right) = -\frac{d}{dz} \left(\rho C_p \overline{\theta'w'} \right). \tag{3.18}$$

In the very idealized model depicted in Figure 3.4, $\partial \overline{\theta'w'}/\partial z$ is independent of height, so Equation 3.18 becomes:

$$\frac{d\bar{\theta}}{dt} = \frac{\left(\overline{\theta'w'} \right)_0 - \left(\overline{\theta'w'} \right)_i}{h}. \tag{3.19}$$

Here $\bar{\theta}$ must be taken as the temperature, averaged vertically over the entire boundary layer. Notice that the boundary layer is heated from below by surface layer heat fluxes, and from above by inversion layer heat fluxes. We must now develop equations for the temporal evolution of h, $\bar{\theta}$ and $\Delta\theta$ the potential temperature step at the inversion base. The key to this is that an expression for $\left(\overline{\theta'w'} \right)_i$, the (downward) sensible heat flux at the inversion base, can easily be found by considering that the depth of the boundary layer h grows at a rate dh/dt, and this

---

[1] To be strictly correct, $\rho C_p \overline{\theta'w'}$ must be called the turbulent sensible heat flux density since it has units W m^{-2}, while $\overline{\theta'w'}$ has units K m s^{-1}.

Figure 3.3 A slant image of a polluted convective boundary layer. The nearly horizontal cloud band marks the base of the overlying inversion layer. The cloud top is gently undulating, with very little vertical structure due to the strongly stable air overlying the polluted mixed layer below the cloud. Pollutants emitted from near the surface are effectively mixed upward through the layer, but are unable to penetrate into the stable layer above. In this image, the polluted layer is about 900 m deep. Photo courtesy of Stephen Oldroyd.

involves the upward movement of the temperature step $\Delta\theta$. In these conditions, the enthalpy gain by the boundary layer must equal the enthalpy loss by the inversion layer. In kinematic terms:

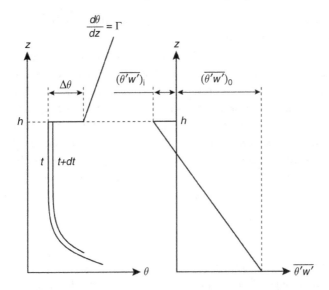

Figure 3.4 Idealized profiles of potential temperature and sensible heat fluxes in a convectively dominated atmospheric boundary layer. h is the depth of the convective boundary layer, $\Delta\theta$ is the temperature step at the inversion base. Potential temperature profiles are indicated at times t and $t + \Delta t$. $\overline{w'\theta'}$ is the kinematic flux of sensible heat. Subscripts 0 and i indicate surface and inversion base respectively.

$$-\left(\overline{\theta'w'}\right)_i = \Delta\theta \frac{dh}{dt}. \quad (3.20)$$

Examination of Figure 3.4 shows that $\Delta\theta$ will decrease as the boundary layer warms (θ decreases) but will increase as the boundary layer gets deeper. If we assume that the lapse rate in the capping inversion ($\Gamma = d\theta/dz$) is constant, then changes in $\Delta\theta$ are given by:

$$\frac{d\Delta\theta}{dt} = \Gamma\frac{dh}{dt} - \frac{d\bar{\theta}}{dt}. \quad (3.21)$$

The assumption of constant Γ is reasonable because it is determined by large scale atmospheric processes that change only slowly compared to processes driven by boundary layer turbulence. Equations 3.19 and 3.21 taken together lead to:

$$\frac{d\Delta\theta}{dt} = \Gamma\frac{dh}{dt} - \frac{\left(\overline{\theta'w'}\right)_0 - \left(\overline{\theta'w'}\right)_i}{h}. \quad (3.22)$$

Equations 3.19, 3.20 and 3.22 are the desired system governing the behaviour of $\bar{\theta}$, h and $\Delta\theta$.

There remains a difficulty. The external variable Γ can be measured by free-floating balloons or aircraft, and $\left(\overline{\theta'w'}\right)_0$ can be measured or estimated using fast response anemometers and thermometers in the surface layer. However, the quantities $\left(\overline{\theta'w'}\right)_i$ and $\Delta\theta$ are internal to the problem in the sense that they are determined by external forcing, rather than being external forcings themselves. They are inaccessible to measurement, yet play crucial roles in governing the convective boundary layer depth. What is needed is a way of determining one of them. An examination of Figure 3.4 reveals that the simple parameterization for $\Delta\theta$:

$$\left(\overline{\theta'w'}\right)_i = -A\left(\overline{\theta'w'}\right)_0 \tag{3.23}$$

is a consequence of the profile geometry. This formulation is supported by a wide range of observations of mixed layer profiles, and can also be arrived at by a consideration of the budget of turbulent kinetic energy at the inversion base, where energy provided by turbulence from below does work drawing air down from the statically stable inversion layer. The energy budget approach provides an estimate $A = 0.20$. For the moment we will retain A as an unknown parameter. This parameterization provides considerable simplification.

Equations 3.20, 3.22 and 3.23 can be combined to give

$$\frac{d\Delta\theta}{dh} + \left(\frac{1+A}{A}\right)\frac{\Delta\theta}{h} - \Gamma = 0, \tag{3.24}$$

which is readily integrated to yield

$$\Delta\theta = \left(\frac{A}{1+2A}\right)\Gamma h.$$

Using this to eliminate $\Delta\theta$ from Equation 3.22 gives

$$\frac{dh}{dt} = \frac{\left(\overline{\theta'w'}\right)_0(1+2A)}{\Gamma h},$$

which integrates to:

$$h^2(t) = h^2(t_0) + \frac{2(1+2A)}{\Gamma}\int_{t_0}^{t}\left(\overline{\theta'w'}\right)_0 d\tau. \tag{3.25}$$

Similar analyses lead to:

$$\Delta\theta(t) = \frac{A\Gamma}{(1+2A)}h(t),$$

and

$$\bar{\theta}(t) = \bar{\theta}(t_0) + \Gamma\frac{(1+A)}{(1+2A)}h(t).$$

A relatively simple generalization of the analysis can be used to accommodate temporally varying Γ.

In these solutions it is assumed that the boundary layer has an initial depth $h(t_0)$ and temperature $\bar{\theta}(t_0)$ at time t_0, which would be the time of day when the surface layer sensible heat flux density $\overline{(\theta'w')}_0$ first becomes positive (upward) due to solar heating of Earth's surface. Under clear sky conditions, this will usually be around or shortly after sunrise. Initial conditions $h(t_0)$ and $\theta(t_0)$, and forcing quantities $\overline{(\theta'w')}_0(t)$ and Γ must be provided by measurements. Such analyses have often been done, and they generally confirm that $A = 0.20$ as a mean value. More detailed analyses of the problem indicate that A does vary slightly through the day, and this is confirmed by measurement programmes.

For mid latitudes, near the equinox, a very close approximation for the sensible heat flux forcing is

$$\overline{(\theta'w')}_0(t) = H\sin(\omega(t-t_0)),$$

and Equation 3.25 integrates to:

$$h^2(t) = h^2(t_0) + \frac{2(1+2A)H}{\Gamma\omega}(1 - \cos(\omega(t-t_0))). \tag{3.26}$$

Useful values are $H = 0.2$ K m s^{-1}, $\Gamma = 0.005$ K m^{-1}, and $\omega = \pi/12\,h^{-1}$. Assuming $h(t_0) = 200$ m at 0600, this simple model results in a convective boundary layer evolution shown in Figure 3.5. The maximum afternoon convective boundary layer depth of about 1800 m is typically what is observed over flat land over continental interiors. This supports the general features of the model, as well as the value $A = 0.2$.

It is worthwhile considering here what has been achieved in our analysis of the convective boundary layer depth. The starting equations were simply conservation of enthalpy, applied to a convectively mixed boundary layer. The crucial step was a parameterization of the

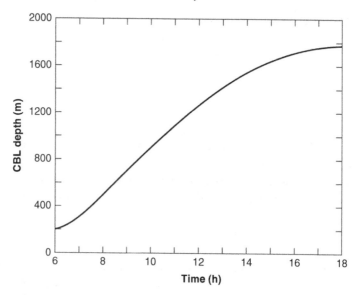

Figure 3.5 Modelled evolution of convective boundary layer depth from Equation 3.26 using the specified forcings and a simple sinusoidal heat flux.

(downward) heat flux at the inversion base as a linear function of the surface layer (upward) heat flux. This simple parameterization implies an infinitesimally thin inversion layer with a finite temperature step, moving upwards at the entrainment velocity. This simple model for entrainment is highly successful at modelling the mixed layer depth. More detailed analyses can be undertaken to yield models that have entrainment layers of finite depth, and rather more detailed parameterizations of the entrainment heat flux. While the more complicated models do provide considerable insight into the physics of entrainment, they give no more information about the diurnal evolution of mixed layer depth than the present simple model.

Further advances in the study of adiabatic and convectively dominated layers, as well as intermediate cases of boundary layer dynamics, are properly undertaken through the application of LES numerical models. These models, when properly evaluated against observations, give access to enormously detailed studies of a wide range of characteristics of all boundary layers. Simple analytical models such as the three presented here are essential benchmarks against which numerical modelling studies are compared, and always form the starting point for numerical modelling studies. The outputs of properly evaluated numerical models can be treated as if they were data, leading to the possibility of statistical or dynamical analyses of forcing processes within the numerical model, including the detailed physics of entrainment processes. Such numerical modelling studies are far easier and cheaper to conduct than field, observational studies, and allow the possibility of manipulating boundary and forcing conditions at will, something that is not possible in measurement studies. There exists a highly influential, but quite small, set of tank model studies of convectively dominated boundary layers that have shed great light on entrainment processes. In a very real way, the science of atmospheric boundary layer physics relies equally on analytical models, numerical models, field observational studies and, to a limited extent, tank studies.

3.3 Sea breezes

Among the many atmospheric dynamical phenomena, the *sea breeze* is fascinating because of its dynamics, important because of its effects on humans and the environment, and pleasant because of its coastal location and gentle nature. As can be seen from Figure 1.1, sea breezes are central in the mesoscale, having a vertical dimension of 1 to 2 km, a horizontal dimension of about 100 km and a typical mean horizontal wind speed of up to 6 m s^{-1}. Notably, the sea breeze draws its driving power from local onshore directed pressure gradients that have their origin in local temperature gradients. Because the sea breeze does not depend on pressure gradients from larger scale phenomena, it can exist in energetic isolation from the surrounding atmosphere, and can therefore be treated as an isolated phenomenon. Of course in many cases, sea breezes are imbedded in larger scale flows which cannot be ignored, but rather must be treated as non-zero boundary conditions.

Temperature gradients that drive sea breezes arise because during daytime, solar energy absorbed by land surfaces produces much greater heating than the same amounts of energy absorbed by water surfaces. This is because water has greater heat capacity than land, water conducts heat more readily than land, evaporative cooling is greater over water than land, and ocean currents and turbulent mixing can

Figure 3.6 A satellite image of a sea breeze front moving westward across the Kenyan coast. The front is marked by the line of clouds west of a zone of cloud-free air. In this image the front is about 50 km from the coast. Three hours later it had advanced to 60–90 km inland. Image courtesy of EuMetSat.

redistribute the heat much more effectively than conduction can over land. A few moments' consideration will soon lead to the conclusion that sea breezes will have a reverse direction by nighttime, since the long-wave radiative loss by land and sea will reverse the driving temperature gradient. This means that sea breezes are onshore winds by daytime, and have an offshore counterpart wind at night, called a *land breeze*. Notice here that we have used the meteorological convention in which a wind is named after the direction from which it comes. A directly observable consequence of this is alternation between a daytime onshore sea breeze and a nighttime offshore land breeze. Also notable is the often sharp reduction in temperature and increase in humidity as air from over water surfaces is carried inland by the sea breeze. These changes can be so sharp that they are termed the *sea breeze front*, as shown in Figure 3.6. The formation of the onshore pressure gradient

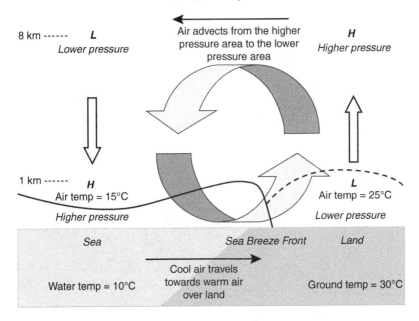

Figure 3.7 Schematic vertical cross-section through a well-developed sea breeze circulation.

can be understood by realizing that the differential heating will result in a greater deepening of the lower atmosphere over land than over sea. While surface pressures will remain the same (because the total column of atmosphere remains unchanged) pressures about a kilometre aloft over land will be greater than over sea because of the differential deepening. This will result in an offshore pressure gradient aloft, which drives a weak offshore flow aloft. This flow redistributes atmospheric mass from land to over sea, producing a low-level, onshore pressure gradient which drives the sea breeze. As a consequence, the sea breeze and its corresponding return flow aloft form a two-dimensional circulation, centred over the coastline. Many of these features are captured diagramatically in Figure 3.7.

Using the scales noted above, the Rossby number for sea breezes is:

$$Ro \leq \frac{10}{10^{-4}10^5} = 1,$$

and from our earlier discussion, we conclude that the Coriolis force will be important for sea breezes. If the Coriolis force dominates, sea

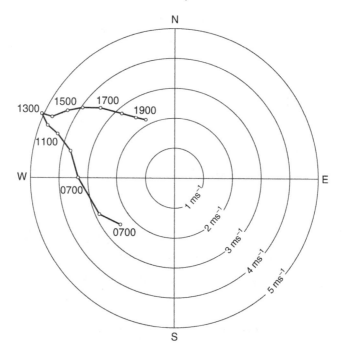

Figure 3.8   Average diurnal sea breeze hodograph (based on 40 independent cases) at Logan Airport, Boston, MA, USA. Based on Haurwitz, B., 1947: Comments on the sea breeze circulation. *Journal of Meteorology*, **40** (1), 1–8.

breeze direction should rotate due to the Coriolis effect, and the rotation should be clockwise in the Northern Hemisphere and anticlockwise in the Southern Hemisphere. The alternation of wind direction across the coastline under land and sea breezes thus becomes a steady rotation, rather than a simple 180° degree reversal of direction.

The conventional graphical device for representing the dependence of vector wind at a single location on some other variable is called a hodograph. Most commonly, the independent variable is either time or height. In studies of sea breezes, a temporal hodograph is used.

Figure 3.8 illustrates such a hodograph, showing that during daytime, the wind rotates steadily in a clockwise direction from 3 m s^{-1} from the WSW at 0700 to 3 m s^{-1} NNE at 2000, and reaches a maximum speed of 5 m s^{-1} at 1300. Data for nighttime land breezes were not available. The most salient feature of this hodograph is the systematic clockwise rotation through the day. When nighttime wind data are added to the hodograph, it is seen that the rotation is a feature of the aforementioned daily alternation of wind direction across the coastline. It is tempting simply to assume that the clockwise rotation is due to dominance of the Coriolis effect, but there exist observations of sea breezes in the Northern Hemisphere that exhibit anticlockwise rotation. Clearly the hodograph rotation is an essential feature of the sea breeze horizontal wind component, and should be amenable to analysis using the horizontal momentum equations, Equations 1.9a and 1.9b. Such an analysis should reveal the balance of forces underlying the dynamics of hodograph rotation. A simple approach to this analysis was provided by Bernhard Haurwitz, one of the founders of modern, analytical dynamic meteorology. I present his 1947 simple linear analytical model of the sea breeze hodograph rotation.

The horizontal momentum equations, Equations 1.9a and 1.9b, are:

$$\frac{du}{dt} = fv - \frac{1}{\rho}\frac{\partial p}{\partial x} + v\left(\frac{\partial^2 u}{\partial x^2} + \frac{\partial^2 u}{\partial y^2} + \frac{\partial^2 u}{\partial z^2}\right)$$
$$\frac{dv}{dt} = -fu - \frac{1}{\rho}\frac{\partial p}{\partial y} + v\left(\frac{\partial^2 v}{\partial x^2} + \frac{\partial^2 v}{\partial y^2} + \frac{\partial^2 v}{\partial z^2}\right). \tag{3.27}$$

There will clearly be a fairly complex interaction between the three accelerating terms on the right hand sides of these equations, and of course the friction terms remain problematical. Also difficult is how to accommodate the pressure effect due to land–sea temperature differences. Haurwitz dealt with these difficulties by strong linearizations of the equations. The effect of friction is assumed to be represented by a decelerating force that depends linearly on wind speed. This is a rather dramatic linearization, since it is known that a more realistic form has friction proportional to the square of wind speed. The justification for the linear assumption is the narrow range and modest magnitude of wind speeds encountered in sea breezes. Thus the last terms on the right hand sides of Equation 3.27 are replaced by $-Ku$ and $-Kv$ respectively. Further, Haurwitz assumes that the pressure gradient term in Equation 3.27 can be replaced by the conventional pressure gradient and a pressure force term $G(t)$ due to the (horizontal) land–sea temperature differences.

We use a coordinate system in which the coastline is straight, and parallel to the y axis. Since the pressure force due to horizontal temperature differences is oriented perpendicular to the coastline, $G(t)$ has only an x component. Onshore and cross-shore wind components and coordinates are given by u, v, and x, y respectively. Equation 3.27 thus becomes:

$$\frac{du}{dt} = fv - \frac{1}{\rho}\frac{\partial p}{\partial x} - G(t) - Ku$$
$$\frac{dv}{dt} = -fu - \frac{1}{\rho}\frac{\partial p}{\partial y} - Kv. \tag{3.28}$$

Since the geostrophic wind is an external condition, and assumed constant over the lifetime of a single sea breeze cycle, we replace

$$-\frac{1}{\rho}\frac{\partial p}{\partial x} \quad \text{by} \quad G_x, \quad \text{and} \quad -\frac{1}{\rho}\frac{\partial p}{\partial y} \quad \text{by} \quad G_y,$$

which leads to:

$$\frac{du}{dt} - fv + Ku = G_x - G(t)$$
$$\frac{dv}{dt} + fu + Kv = G_y. \tag{3.29}$$

$G(t)$ must reflect the alternation of thermal forcing across the coastline. A convenient way of doing this is to specify:

$$G(t) = \frac{G_0}{\pi} + \frac{G_0}{2}\cos(\omega t),$$

where ω is the diurnal period. G_0 can be estimated by consideration of a height z well above the sea breeze, at which the pressure p is uniform. The Poisson equation 1.10 gives:

$$p_s = p\,\exp\left(\frac{gz}{\mathscr{R}T}\right).$$

Logarithmic differentiation with respect to x, and noting that $\partial p/\partial x = 0$, gives:

$$\frac{1}{p_s}\frac{\partial p_s}{\partial x} = -\frac{gz}{\mathscr{R}T^2}\frac{\partial T}{\partial x},$$

and

$$G_0 \equiv \frac{1}{\rho}\frac{\partial p_s}{\partial x} = \frac{gz}{T}\frac{\partial T}{\partial x}.$$

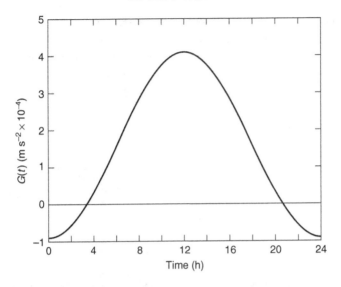

Figure 3.9 Sea breeze thermal forcing function $G(t)$. Based on Haurwitz, B., 1947: Comments on the sea breeze circulation. *Journal of Meteorology*, **40** (1), 1–8.

Observations give the horizontal temperature gradient under sea breezes to be about 5 K in 100 km, and the full sea breeze depth to be about 600 m. If the vertical gradient of temperature is fully expressed over half this height, then $G_0 = 5 \times 10^{-4}$ m s^{-2}.

Figure 3.9 shows $G(t)$ for a mid-latitude location in midsummer.

If we define the complex variables:

$$W = u + iv, \quad G_z = G_x + iG_y,$$

Equation 3.28 becomes

$$\frac{dW}{dt} + (K + if)W = G_z - G(t),$$

which can easily be integrated to yield

$$W = \frac{G_z}{K+if} - \frac{G_0}{\pi} \frac{1}{K+if} - \frac{G_0}{2} \frac{\omega \sin \omega t + (K+if)\cos \omega t}{(K+if)^2 + \omega^2}. \quad (3.30)$$

The constant of integration must be zero since the sea breeze wind must vanish when both pressure gradient terms, G_z and G_0, are zero. The first term on the right hand side of Equation 3.30 is identified as wind driven by the constant part of the pressure gradient force. If there were no friction, $K = 0$, the solution would collapse to the geostrophic wind. The second and third terms represent the effect of pressure gradients which arise from horizontal (land–sea) temperature differences.

Some rearrangement is needed to extract the two velocity components from Equation 3.30. They are:

$$u = \frac{KG_x + fG_y}{K^2 + f^2} - \frac{G_0}{\pi} \frac{K}{K^2 + f^2}$$
$$- \frac{G_0}{2} \frac{(\omega^2 + K^2 - f^2)\omega \sin \omega t + (\omega^2 + K^2 + f^2)K \cos \omega t}{(\omega^2 + K^2 - f^2)^2 + 4K^2 f^2}$$
$$v = -\frac{fG_x - kG_y}{K^2 + f^2} + \frac{G_0}{\pi} \frac{f}{K^2 + f^2} \qquad (3.31)$$
$$- \frac{G_0}{2} \frac{(\omega^2 - K^2 - f^2)f \cos \omega t - 2Kf\omega \sin \omega t}{(\omega^2 + K^2 - f^2)^2 + 4K^2 f^2}.$$

The hodograph representing the diurnal evolution of these wind components is plotted in Figure 3.10, assuming a zero geostrophic wind. In this exercise, a value for K which gives roughly the same magnitude of maximum winds as observed at noon (see Figure 3.8) is used to facilitate a comparison. A value of $K = 0.58 \times 10^{-4}$ s^{-1} is appropriate. A simple comparison of the shape and evolution of the hodograph shows that the model nicely captures the essential features of the sea breeze diurnal evolution. The close correspondence between the observed and modelled hodographs means that the average geostrophic wind on the 40 days used to construct Figure 3.8 is roughly zero.

The effect of non-zero geostrophic wind on the hodograph is seen in Figure 3.11. In this case a geostrophic wind of 5 m s^{-1} from the Southwest is assumed. As can be seen, the hodograph maintains its elliptical shape, but is displaced to the Northeast. It is also clear that, while the sea breeze modelled in Figure 3.10 exhibits continuous clockwise rotation, the sea breeze modelled in Figure 3.11 exhibits both clockwise and anticlockwise rotation, when viewed from the perspective of a point fixed in space (as at a measurement site). This is an indication that sea breeze hodograph rotation is not simply dominated by Coriolis effects, but that other factors come into play.

One final useful result can be drawn from this simple sea breeze model. The influence of friction can be investigated by setting $K = 0$ in Equation 3.31. Doing this results in:

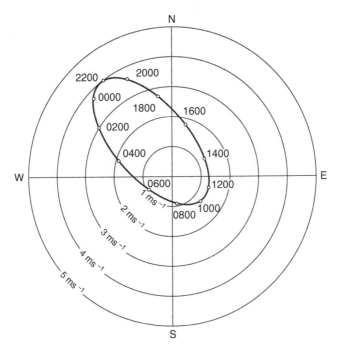

Figure 3.10 Modelled sea breeze hodograph for zero geostrophic wind.
Based on Haurwitz, B., 1947: Comments on the sea breeze circulation.
Journal of Meteorology, **40** (1), 1–8.

$$u = \frac{G_y}{f} - \frac{G_0}{2}\frac{\omega}{(\omega^2 - f^2)}\sin \omega t,$$

$$v = -\frac{G_x}{f} + \frac{G_0}{\pi f} - \frac{G_0}{2}\frac{f}{(\omega^2 - f^2)}\cos \omega t.$$

These equations mean that the sea breeze wind speed will become infinite when $\omega = f$, at latitudes 30° North and South. This is clearly unphysical. Furthermore, even if the sea breeze merely became extremely strong, its effect would be to eliminate all land–sea temperature differences that drive it. This is a form of resonance because the period of inertial oscillations f becomes equal to the periodic thermal forcing ω, and clearly the resonance is kept in control by friction. We can therefore conclude that surface friction plays an essential role in governing the diurnal evolution of the sea breeze at all latitudes. This is as far as the simple linear sea breeze model takes us. There have been

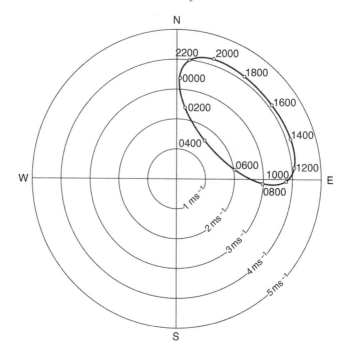

Figure 3.11 Modelled sea breeze hodograph for a geostrophic wind of 5 m s^{-1}. Based on Haurwitz, B., 1947: Comments on the sea breeze circulation. *Journal of Meteorology*, **40** (1), 1–8.

many attempts to build non-linear analytical models of sea breezes, but all eventually result in equations that require numerical treatment. It is therefore not surprising that all recent studies of sea breezes employ numerical models.

Specifically, sea breezes are modelled using any of the many available mesoscale atmospheric numerical models. All these models are fully non-linear and three-dimensional. They contain components which model the physics of atmospheric boundary layers with various levels of parameterized turbulence transport schemes. They are able to accommodate fully realistic topography and coastlines, and can be imbedded in larger scale models that capture dynamically changing larger scale weather conditions. Such modelling exercises can be used to simulate actual sea breeze conditions in realistic locations, and can also be used in idealized simulations to investigate the internal dynamics of sea breezes. These latter exercises can be (and in many cases have been) used to study the effects of realistic friction, realistic heating profiles, three-dimensional structure, coastline curvature, non-flat topography, non-linear advection, overlying atmosphere thermodynamics, sea breeze front initiation mechanism, and a host of outstanding research questions.

One interesting numerical modelling exercise that will illustrate this is the study of sea breeze hodograph rotation on and around the island of Sardinia conducted by Moisseeva and Steyn, using the Weather Research and Forecasting (WRF)[2] model. The model is implemented over a domain covering Sardinia, with realistic topography and surface land-use. The model was run on two nested grids, with a grid resolution of 3 km in the inner grid. Model boundary conditions were supplied from actual meteorological fields, and output was compared with observations at twelve coastal meteorological stations. This comparison was used to show that the model realistically simulated observed diurnal hodograph rotation.

Figure 3.12 shows the remarkably close agreement between WRF model output and station observations of onshore wind speed during one day in the study. Notice that these plots are one-dimensional versions of the sea breeze hodographs displayed in Figures 3.10 and 3.11. The WRF model output is available for every grid square in the modelled domain, and so allows access to hodographs over the entire domain, including land and sea surfaces. More importantly, extracting individual force terms from the horizontal momentum equations at each model time step allows Moisseeva and Steyn to perform a dynamical analysis of the diurnal evolution of wind velocity. Analysis of force balance underlying the winds plotted in Figure 3.12 shows that the direction of hodograph rotation is a result of a complex interaction between near-surface and synoptic pressure gradient, Coriolis and advection forces, and that this balance varies across the domain of study. The research provides new insights into the dynamics underlying sea breeze hodograph rotation, especially in coastal zones with complex topography and/or coastline.

This study exemplifies the use of a numerical model in an instance where non-linear effects are so important that a simple linear model

---

[2] A mesoscale numerical weather prediction system designed for both forecasting and research purposes available through: www.wrf-model.org/index.php.

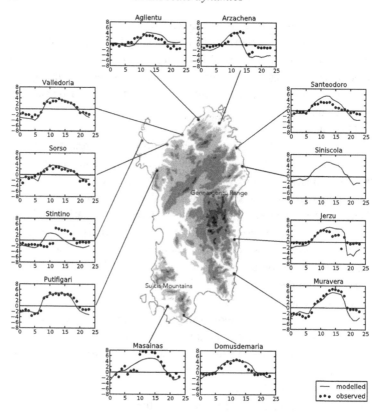

Figure 3.12 Diurnal evolution of modelled and observed onshore wind speed at twelve coastal meteorological measurement stations around Sardinia on 21 June 1998. Grey shading shows smoothed topography. Reproduced from Moisseeva, N. and Steyn, D. G., 2014: Dynamic analysis of sea breeze hodograph rotation in Sardinia. *Atmospheric Chemistry and Physics*, **14**, 13471–13481. doi:10.5194/acp-14-13471-2014, under Creative Commons License http://creativecommons.org/licenses/by/3.0/.

such as that employed by Haurwitz is insufficient. It also illustrates the importance of using observations of the phenomenon in a model evaluation exercise designed to demonstrate that the model captures reality, and can be used to interpret underlying dynamical processes.

4

Large scale dynamics

We now return to the large scale atmosphere to investigate the possibility of approximate regimes in which the governing equations can be simplified. We cannot reasonably expect to find analytical solutions, but rather hope to reveal generally useful characteristics of atmospheric flow through an examination of flow regimes captured by approximate subsets of the full equations. Before we can approach large scale phenomena, it is necessary to develop a set of concepts that are important to large scales of atmospheric flow dynamics. Not incidentally, the ideas to be developed are also extremely useful in analysing atmospheric data.

4.1 Height, pressure and the geopotential

Because the large scale atmosphere is so strongly stratified, and the pressure and density vary strongly in the vertical, it should not be surprising that a careful consideration of the most appropriate vertical coordinate for both analysis and data processing is in order. The objective here is to use our thermodynamic analysis of atmospheric vertical structure to develop a well-behaved height coordinate for analysis of atmospheric motion. Recalling the hydrostatic balance from Chapter 1, we define the *geopotential* at height z, $\Phi(z)$ as the work done in raising a unit mass of atmosphere to height z from mean sea level:

$$\Phi(z) \equiv \int_0^z g \, dz.$$

If we rewrite the equation of state for dry air as $\alpha = \mathscr{R}T/p$, where α is the specific volume of dry air, and note that $d\Phi = g \, dz$, then the rearranged hydrostatic equation $g \, dz = -dp/\rho$ becomes:

$$d\Phi = -\frac{\mathscr{R}T}{p}\,dp = -\mathscr{R}T\,d\ln p.$$

This integrates to a form of the hypsometric equation:

$$\Phi(z_2) - \Phi(z_1) = \mathscr{R}\int_{p_2}^{p_1} T\,d\ln p.$$

It is often convenient to replace $\Phi(z)$ by the *geopotential height*, defined as

$$Z \equiv \frac{\Phi(z)}{g_0}$$

where g_0 is the global average of gravity at mean sea level, $g_0 \equiv 9.80665$ m s^{-2}. Z is numerically almost identical to the geometric height in the lower atmosphere. The hypsometric equation then becomes:

$$\Delta Z = Z_2 - Z_1 = \frac{\mathscr{R}}{g_0}\int_{p_2}^{p_1} T\,d\ln p,$$

and ΔZ is the thickness of atmosphere between pressure levels p_1 and p_2.

It is now clear that, within a single vertical column of the atmosphere, there exists a single-valued, monotonic relationship between pressure and height, suggesting the use of pressure as a vertical coordinate. Noting the identity

$$\left(\frac{\partial p}{\partial x}\right)_z = -\left(\frac{\partial p}{\partial z}\right)_x\left(\frac{\partial z}{\partial x}\right)_p,$$

and substituting for dp/dz from the hydrostatic equation yields:

$$-\frac{1}{\rho}\left(\frac{\partial p}{\partial x}\right)_z = -g\left(\frac{\partial z}{\partial x}\right)_p = -\left(\frac{\partial \Phi}{\partial x}\right)_p.$$

Similarly:

$$-\frac{1}{\rho}\left(\frac{\partial p}{\partial y}\right)_z = -\left(\frac{\partial \Phi}{\partial y}\right)_p.$$

This shows that in an *isobaric* coordinate system, horizontal pressure gradient force is captured by the gradient of geopotential along a surface of constant pressure, and density no longer appears explicitly in the pressure gradient force expression. This constitutes a very helpful linearization of the governing equations. We are now in a position to consider some elementary applications of the equations of motion.

4.2 Geostrophic dynamics

As a starting point in our study of large scale dynamics, we return to geostrophic flow. In Chapter 2 we simplified the governing equations to arrive at the very simple geostrophic wind equations, applicable under conditions in which the Coriolis force dominates all inertial and frictional forces, and is balanced by the pressure gradient force, and $|\rho'/\rho_0| \ll 1$. The equations of motion reduce to:

$$-fv = -\frac{1}{\rho}\frac{\partial p}{\partial x} \tag{4.1a}$$

$$fu = -\frac{1}{\rho}\frac{\partial p}{\partial y} \tag{4.1b}$$

$$0 = -\frac{1}{\rho}\frac{\partial p}{\partial z} \tag{4.1c}$$

$$\frac{\partial u}{\partial x} + \frac{\partial v}{\partial y} + \frac{\partial w}{\partial z} = 0. \tag{4.1d}$$

Taking the vertical derivative of Equation 4.1a gives:

$$-f\frac{\partial v}{\partial z} = -\frac{1}{\rho}\frac{\partial}{\partial z}\left(\frac{\partial v}{\partial x}\right) = -\frac{1}{\rho}\frac{\partial}{\partial x}\left(\frac{\partial v}{\partial z}\right) = 0,$$

where the pressure terms are zero by virtue of Equation 4.1c. A similar analysis shows that the same holds for the other horizontal velocity component. So, in geostrophic atmospheres,

$$\frac{\partial v}{\partial z} = \frac{\partial u}{\partial z} = 0. \tag{4.2}$$

This is quite surprising as it means that air parcels retain their relative vertical position, and the fluid moves like a quasi-rigid body. Any air parcel that is exactly above another always stays exactly above its partner, all the way through the layer. This can be interpreted as a vertical rigidity, and is a fundamental property of rotating homogeneous fluids. As was seen in Chapter 2, the geostrophic wind is always directed along *isobars*, lines of constant pressure. Furthermore, if the flow is over a limited range of latitude so that f is constant, there is a frame of reference called the *f-plane*. In this frame, the horizontal divergence is:

$$\frac{\partial u}{\partial x} + \frac{\partial v}{\partial y} = -\frac{\partial}{\partial x}\left(\frac{1}{\rho f}\frac{\partial p}{\partial y}\right) + \frac{\partial}{\partial y}\left(\frac{1}{\rho f}\frac{\partial p}{\partial x}\right) = 0.$$

Because of this result, and Equation 4.1c, geostrophic flows are there-fore fully non-divergent in the f-plane. If a geostrophic atmosphere flows over a flat surface (ocean or land), the vertical velocity must vanish, and the flow is strictly two-dimensional. This is an enormous simplification to the governing equations.

4.3 Non-geostrophic dynamics and the shallow water model

In Section 2.2.1 we relaxed the condition which made inertial forces negligible when compared with the Coriolis force, and saw that this leads to a definition of the Rossby number. We now return to the analysis and include the inertial forces, but retain the conditions of homogeneity and ignore friction. The applicable equations for the horizontal components of momentum are:

$$\frac{\partial u}{\partial t} + u\frac{\partial u}{\partial x} + v\frac{\partial u}{\partial y} + w\frac{\partial u}{\partial z} - fv = -\frac{1}{\rho}\frac{\partial p}{\partial x}$$
$$\frac{\partial v}{\partial t} + u\frac{\partial v}{\partial x} + v\frac{\partial v}{\partial y} + w\frac{\partial v}{\partial z} + fu = -\frac{1}{\rho}\frac{\partial p}{\partial y}.$$

$$(4.3)$$

If the atmosphere starts in a state of purely geostrophic flow, $u\partial u/\partial x$ and similar terms are all independent of z. Furthermore, the Coriolis term is z independent, and from Equation 4.1c, the pressure is z inde-pendent. The result is that $\partial u/\partial t$ and $\partial v/\partial t$ are also independent of z, and u and v are independent of z for all time:

$$\frac{\partial u}{\partial t} + u\frac{\partial u}{\partial x} + v\frac{\partial u}{\partial y} - fv = -\frac{1}{\rho}\frac{\partial p}{\partial x}$$
$$\frac{\partial v}{\partial t} + u\frac{\partial v}{\partial x} + v\frac{\partial v}{\partial y} + fu = -\frac{1}{\rho}\frac{\partial p}{\partial y}.$$

$$(4.4)$$

Atmospheric flows obeying these equations are called *barotropic*, and constitute an important special case of atmospheric flow. These flows have no vertical structure, but unlike geostrophic flows, the flow does not follow isobars, and can have non-zero vertical velocity. Their ver-tical velocity can be determined by an examination of the continuity equation:

$$\frac{\partial u}{\partial x} + \frac{\partial v}{\partial y} + \frac{\partial w}{\partial z} = 0,$$

in which the first two terms are z independent, and do not necessarily add up to zero. A vertical velocity that is linear in z is permitted by the equations, and this will lead to non-zero horizontal divergence, and an associated flow across isobars. If we imagine that the atmosphere has a well-defined depth $h(x,y,t)$, then the continuity equation can be integrated over h to yield:

$$\left(\frac{\partial u}{\partial x}+\frac{\partial v}{\partial y}\right)\int_0^h dz+[w]_0^h=0. \tag{4.5}$$

Since the atmosphere is restricted to remain between 0 and h, and we are assuming a flat underlying surface, vertical velocities at these levels are given by:

$$w(z=h)=\frac{\partial h}{\partial t}+u\frac{\partial h}{\partial x}+v\frac{\partial h}{\partial y}$$

$$w(z=0)=0.$$

Using the upper and lower limits on w, Equation 4.5 becomes:

$$\frac{\partial h}{\partial t}+\frac{\partial hu}{\partial x}+\frac{\partial hv}{\partial y}=0.$$

Since the atmosphere is assumed homogeneous, the dynamic pressure is $p=\rho gh$, and z disappears from the governing equations which become:

$$\frac{\partial u}{\partial t}+u\frac{\partial u}{\partial x}+v\frac{\partial u}{\partial y}-fv=-g\frac{\partial h}{\partial x} \tag{4.6a}$$

$$\frac{\partial v}{\partial t}+u\frac{\partial v}{\partial x}+v\frac{\partial v}{\partial y}+fu=-g\frac{\partial h}{\partial y} \tag{4.6b}$$

$$\frac{\partial h}{\partial t}+\frac{\partial hu}{\partial x}+\frac{\partial hv}{\partial y}=0. \tag{4.6c}$$

These equations can easily be generalized to a non-flat bottom, with height $b(x,y)$ above some reference level, and are very useful for understanding large scale flow of the lower atmosphere. Because they were originally derived for the ocean, they are called the *shallow water equations*. The equations remain non-linear, and therefore do not yield to analytical solutions, except under very restrictive boundary conditions and strong linearization. One set of solutions corresponds to geostrophic flow, another produces gravity-inertia waves. Attempts at solutions with non-flat (even idealized) topography must eventually rely on numerical approaches.

4.4 Vertical shear of the geostrophic wind

It is commonly observed that there exist horizontal gradients of temperature at large scales, generally due to the juxtaposition of cold and warm air masses, as depicted in panel (a) of Figure 4.1. The geostrophic wind is proportional to the horizontal pressure gradient, and if the atmosphere is assumed to be hydrostatic, panel (b) of Figure 4.1 shows that the geostrophic wind (parallel to the y axis) increases with height in response to increasing slope of the isobars. To see this, consider the height increment δz as a function of pressure increment δp. The hydrostatic approximation has:

$$\delta z \approx -\frac{\delta p}{p}\frac{\mathscr{R}T}{g}$$

$$v_g = \frac{1}{f}\frac{\partial \Phi}{\partial x} \quad \text{and} \quad u_g = \frac{1}{f}\frac{\partial \Phi}{\partial y}. \tag{4.7}$$

Note that the derivatives must be taken along surfaces of constant pressure. In these coordinates, using the ideal gas law, the hydrostatic balance is:

$$\frac{\partial \Phi}{\partial p} = -\alpha = -\frac{\mathscr{R}T}{p}. \tag{4.8}$$

Differentiating Equation 4.7 with respect to pressure, and applying Equation 4.8 gives:

$$\begin{aligned}p\frac{\partial v_g}{\partial p} \equiv \frac{\partial v_g}{\partial \ln p} = -\frac{\mathscr{R}}{f}\left(\frac{\partial T}{\partial x}\right)_p \\ p\frac{\partial u_g}{\partial p} \equiv \frac{\partial u_g}{\partial \ln p} = -\frac{\mathscr{R}}{f}\left(\frac{\partial T}{\partial y}\right)_p,\end{aligned} \tag{4.9}$$

which in vector form is

$$\frac{\partial \mathbf{V}_g}{\partial \ln p} = -\frac{\mathscr{R}}{f}\mathbf{k}\times(\nabla T)_p. \tag{4.10}$$

Equation 4.10 is generally referred to as the *thermal wind* equation, but in reality expresses the vertical geostrophic wind shear as a function of horizontal temperature gradients. The thermal wind itself is then

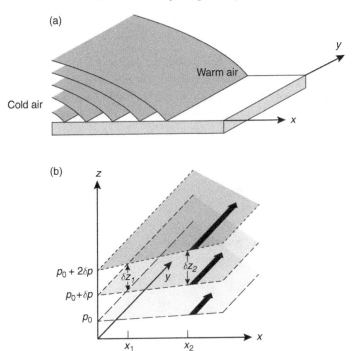

Figure 4.1 Schematic depiction of juxtaposition of: (a) cold air mass (on the left) positioned against a warm air mass. The cold air forms a wedge under warm air because of its greater density. This results in horizontal temperature gradients at all levels, with warmer air in the positive x direction. (b) Detail from panel (a) showing vertical shear of the geostrophic wind arising from horizontal temperature gradients at all levels. Dashed surfaces are isobars. Note that $T_{x_2} > T_{x_1}$ and $\delta p < 0$.

the vector difference between geostrophic winds at two levels. We can quantify it by integrating Equation 4.10 between two pressure levels, p_0 and p_1, where $p_1 < p_0$ because hydrostatic conditions prevail:

$$\mathbf{V}_T = \mathbf{V}_g(p_1) - \mathbf{V}_g(p_0) = -\frac{\mathcal{R}}{f} \int_{p_0}^{p_1} (\mathbf{k} \times \nabla T)\, d\ln p. \qquad (4.11)$$

If the mean temperature of the layer between pressure levels p_0 and p_1 is $\bar{T}$, the thermal wind has components:

$$u_T = -\frac{\mathcal{R}}{f} \left(\frac{\partial \bar{T}}{\partial y}\right)_p \ln\left(\frac{p_0}{p_1}\right) \qquad (4.12)$$

$$v_T = -\frac{\mathscr{R}}{f} \left(\frac{\partial \bar{T}}{\partial x} \right)_p \ln \left(\frac{p_0}{p_1} \right). \tag{4.13}$$

The geopotential coordinate introduced earlier provides a convenient alternative representation of the thermal wind. The thermal wind can be related to the horizontal gradient of the top to bottom geopotential difference for the layer under consideration. To see this, we must integrate Equation 4.8 vertically through the layer from p_0 to p_1 and use the mean layer temperature $\bar{T}$.

This gives:

$$\Phi_1 - \Phi_0 = \mathscr{R}\ln \left(\frac{p_0}{p_1} \right) \bar{T},$$

where $\Phi_1 - \Phi_0$ is the thickness of the layer between pressure levels p_0 and p_1 measured in geopotential units.

This then easily leads to:

$$u_T = u_g(p_1) - u_g(p_0) = -\frac{\mathscr{R}}{f} \frac{\partial}{\partial y} (\Phi_1 - \Phi_0) \tag{4.14}$$

$$v_T = v_g(p_1) - v_g(p_0) = -\frac{\mathscr{R}}{f} \frac{\partial}{\partial x} (\Phi_1 - \Phi_0). \tag{4.15}$$

It is worth noting that lines of equal $\Phi_1 - \Phi_0$ can be easily related to lines of equal mean layer temperature, and are often used for this purpose in plotting weather maps. The thermal wind equation in any of its forms is extremely useful for diagnosing analysed wind and temperature fields, and is also used for estimating the geostrophic wind at some level if the geostrophic wind at some other level and the mean temperature gradient are known. These are all tasks of importance to weather forecasters, particularly when preparing aviation forecasts.

A particular special case is that of a *barotropic atmosphere* in which density depends only on pressure $\rho \equiv \rho(p)$. Here, isobaric surfaces are also constant density surfaces. As the atmosphere behaves as an ideal gas, isobaric surfaces are also isotherms, and $(\nabla T)_p = 0$, and the thermal wind equation, Equation 4.10, becomes:

$$\frac{\partial \mathbf{V}_g}{\partial \ln p} = 0. \tag{4.16}$$

The assumption of barotropic conditions is an enormously restrictive one, and clearly provides strong restrictions on the kind of dynamics that may occur. Analysis of data on the global atmosphere shows that barotropic conditions exist in both equatorial and polar regions. If temperature varies along isobars, the atmosphere is in a *baroclinic* state, one in which the density depends on both pressure and temperature: $\rho \equiv \rho(p, T)$. Vast regions of baroclinicity exist in the mid latitudes, forming a transition zone between the two barotropic regions. The baroclinicity is driven by horizontal temperature gradients between equatorial and sub-polar regions. The mid-latitude baroclinic zones are the location of moving, mid-latitude cyclonic storms that derive their energy from zones of very intense horizontal temperature gradients. These storms intensify in the winter seasons in both hemispheres as equator to pole temperature gradients become steeper.

4.5 Vorticity in the atmosphere

It is commonly known that the mesoscale and largescale atmosphere is populated by many phenomena that have swirling or rotating motion. These phenomena include dust devils, tornadoes, cyclones, typhoons and hurricanes. It should therefore be obvious that special attention must be paid to the quantification of vortex-like motion in the atmosphere.

4.5.1 Vorticity in the shallow water equations

In studying geostrophic flows, we noted that flow was along isobars, meaning that no work is done by or against pressure. This can only occur if pressure terms in expressions for the horizontal divergence either vanish or cancel. If we return to Equations 4.6a and 4.6b, cross differentiate (allowing for the Coriolis force to vary horizontally), subtract and rearrange we get:

$$\frac{d}{dt}\left(f + \frac{\partial v}{\partial x} - \frac{\partial u}{\partial y}\right) + \left(\frac{\partial u}{\partial x} + \frac{\partial v}{\partial y}\right)\left(f + \frac{\partial v}{\partial x} - \frac{\partial u}{\partial y}\right) = 0, \quad (4.17)$$

where the two-dimensional material derivative is

$$\frac{d}{dt} = \frac{\partial u}{\partial t} + u\frac{\partial u}{\partial x} + v\frac{\partial v}{\partial y}.$$

Noting that *vorticity* measures the property of rotation in a fluid, and is defined as the curl of the velocity vector field, and that the vertical component of the curl of vector **V** is

$$\left(\frac{\partial V_y}{\partial x} - \frac{\partial V_x}{\partial y}\right),$$

we can interpret the grouping $(f + \partial v/\partial x - \partial u/\partial y)$ as the sum of background vorticity caused by the Coriolis induced rotation (f) and intrinsic or *relative vorticity* $(\zeta = \partial v/\partial x - \partial u/\partial y)$, and we can write

$$\left(f + \frac{\partial v}{\partial x} - \frac{\partial u}{\partial y}\right) = f + \zeta.$$

Notice that in the present case, the horizontal flow field is independent of z. Since there is no vertical shear, there can be no eddies with horizontal axes, and the vorticity has only a vertical component. Using the preceding logic, Equation 4.6c, the continuity equation, can be expanded to:

$$\frac{dh}{dt} + \left(\frac{\partial u}{\partial x} + \frac{\partial v}{\partial y}\right) h = 0. \qquad (4.18)$$

If we now consider the cross-sectional area ds of a thin fluid column, we can write an equation for the variation in ds as it moves along and is distorted by the flow:

$$\frac{d}{dt} ds = \left(\frac{\partial u}{\partial x} + \frac{\partial v}{\partial y}\right) ds. \qquad (4.19)$$

It is easily seen that horizontal divergence/convergence causes ds to increase/decrease. Equations 4.18 and 4.19 lead to:

$$\frac{d}{dt}(h\,ds) = 0, \qquad (4.20)$$

which means that the volume of the column of air is conserved. If the column increases in depth, it will become more slender. A similar treatment of Equations 4.17 and 4.19 leads to:

$$\frac{d}{dt}[(f + \zeta)\,ds] = 0, \qquad (4.21)$$

which implies that the product $(f + \zeta)\,ds$ for the fluid column is conserved. This condition is closely analogous to the conservation of angular momentum for an isolated rotating rigid body. If both vorticity and volume are conserved, the ratio of these two quantities must also be conserved, so that:

$$\frac{d}{dt}\left(\frac{f + \zeta}{h}\right) = 0. \qquad (4.22)$$

The quantity

$$\frac{f+\zeta}{h} \equiv \frac{f+\frac{\partial v}{\partial x} - \frac{\partial u}{\partial y}}{h}$$

is independent of the column cross-section, and depends only on local flow quantities. It is known as the *potential vorticity*. As we started this analysis with Equation 4.4, the results strictly only apply to barotropic flows, and our definition of potential vorticity only applies to shallow water flows of depth h.

4.5.2 The vorticity equation

The concepts presented in Section 4.5.1 can be usefully applied to a much wider range of flows than barotropic conditions in the shallow water equations. In general, the vertical component of relative vorticity ζ is associated with cyclonic scale storms in the mid latitudes – large positive/negative ζ occurs in such storms in the Northern/Southern Hemisphere, in which strongly baroclinic conditions prevail. We now develop an equation for vertical vorticity that is less limiting than the approach of Section 4.5.1. We return to Equation 2.2, in which we retained the acceleration term in the geostrophic balance:

$$\frac{du}{dt} - fv = -\frac{1}{\rho}\frac{\partial p}{\partial x} \tag{4.23}$$

$$\frac{dv}{dt} + fv = -\frac{1}{\rho}\frac{\partial p}{\partial y}. \tag{4.24}$$

If we take the y derivative of the x component equation, and the x derivative of the y component equation, then subtract the result and use the definition $\zeta = \partial v/\partial x - \partial u/\partial y$, we obtain:

$$\frac{\partial \zeta}{\partial t} + u\frac{\partial \zeta}{\partial x} + v\frac{\partial \zeta}{\partial y} + w\frac{\partial \zeta}{\partial z} + (\zeta + f)\left(\frac{\partial u}{\partial x} + \frac{\partial v}{\partial y}\right)$$
$$+ \left(\frac{\partial w}{\partial x}\frac{\partial v}{\partial z} + \frac{\partial w}{\partial y}\frac{\partial u}{\partial z}\right) + v\frac{df}{dy} = \frac{1}{\rho^2}\left(\frac{\partial \rho}{\partial x}\frac{\partial p}{\partial y} - \frac{\partial \rho}{\partial y}\frac{\partial p}{\partial x}\right). \tag{4.25}$$

Recognizing that the Coriolis parameter depends only on y so that $df/dt = vdf/dy$, we can rearrange Equation 4.25 to obtain the *vorticity equation*:

$$\frac{d}{dt}(\zeta + f) = -(\zeta + f)\left(\frac{\partial u}{\partial x} + \frac{\partial v}{\partial y}\right)$$

$$-\left(\frac{\partial w}{\partial x}\frac{\partial v}{\partial z} - \frac{\partial w}{\partial y}\frac{\partial u}{\partial z}\right)$$

$$+\frac{1}{\rho^2}\left(\frac{\partial \rho}{\partial x}\frac{\partial p}{\partial y} - \frac{\partial \rho}{\partial y}\frac{\partial p}{\partial x}\right). \tag{4.26}$$

As presented above, the rate of change in vorticity arises from three processes, identified with the three terms on the right hand side of Equation 4.26: the divergence term, the tilting term and the solenoidal term.

- Divergence term $(\zeta + f)(\partial u/\partial x + \partial v/\partial y)$: the effects of this term are analogous to the conservation of angular momentum for a rigid body. If the moment of inertia of the body changes (for example when a spinning skater draws in her legs or arms and as a consequence increases her rate of spin), conservation of momentum requires that the angular velocity must change. If the flow is converging, the vortex decreases in horizontal dimension, and thus must increase in vortical velocity.

- Tilting term $(\partial w/\partial x)(\partial v/\partial z) - (\partial w/\partial y)(\partial u/\partial z)$: if the vertical velocity field has horizontal gradients, vertical vorticity will be increased as horizontally oriented vorticity components are tilted into the vertical.

- Solenoidal term $(\partial \rho/\partial x)(\partial p/\partial y) - (\partial \rho/\partial y)(\partial p/\partial x)$: this term represents the vertical vorticity generated if surfaces of constant density do not coincide with surfaces of constant pressure, as illustrated in Figure 4.2. Variations in density produce variations in the pressure gradient force, and variations in acceleration result in the production of vorticity. This term must be zero in barotropic conditions.

It is useful to perform a scale analysis of terms in the vorticity equation, so we return to the set of scales for cyclonic storms presented in Table 2.2.

If the reference Coriolis parameter for mid latitudes is f_0, and we expand the latitude-dependent Coriolis parameter in a Taylor series, neglecting all second order and higher terms, we have:

$$f = f_0 + \beta y + \cdots,$$

where latitude is represented by the North–South Cartesian coordinate y, and the *beta parameter* is the linear latitude dependence of the

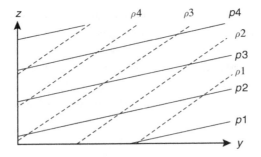

Figure 4.2 Schematic diagram of an x–z plane in a baroclinic atmosphere showing crossing of pressure and density surfaces resulting in non-zero solenoidal term.

Coriolis parameter. This *beta plane approximation*, being linear, renders the equations at least partially tractable.

We must proceed carefully here because terms in the vorticity equation consist of sums of individual terms having the same order of magnitude, but possibly having different signs. The resultant cancellation means that the sum will have an upper limit to its magnitude given by the magnitude of the individual terms. We will represent this by the symbol $\lesssim$, meaning less than or equal to in order of magnitude. For example:

$$\zeta = \frac{\partial v}{\partial x} - \frac{\partial u}{\partial y} \lesssim \frac{U}{L} \sim 10^{-5}\,\mathrm{s}^{-1}.$$

Note that the ratio of relative to background vorticity, $\zeta/f_0 \lesssim U/(f_0 L) \equiv Ro \sim 10^{-1}$. This means that for mid-latitude cyclonic storms, ζ may be neglected compared to f_0, and the divergence term in Equation 4.26 becomes:

$$(\zeta + f)\left(\frac{\partial u}{\partial x} + \frac{\partial v}{\partial y}\right) \approx f\left(\frac{\partial u}{\partial x} + \frac{\partial v}{\partial y}\right).$$

Terms in Equation 4.26 have magnitudes (all in units s^{-2}):

$$\frac{d}{dt}(\zeta + f) \sim \frac{U^2}{L^2} \sim 10^{-10}$$

$$\text{Divergence term} \lesssim \frac{f_0 U}{L} \sim 10^{-9}$$

$$\text{Tilting term} \lesssim \frac{W U}{H L} \sim 10^{-11}$$

$$\text{Solenoidal term} \lesssim \frac{\Delta p \Delta \rho}{\rho^2 L^2} \sim 10^{-11}.$$

It should be clear from this scale analysis that, in order for a balance to exist, the two derivatives in the divergence term must partially cancel, or else there will be an unconstrained production or destruction of vorticity from divergence processes. Motions at this scale must therefore be partially non-divergent, and the dynamics are dominated by details of horizontal divergence. This conclusion leads to an approximate vorticity equation:

$$\frac{d_h}{dt}(\zeta + f) = -f\left(\frac{\partial u}{\partial x} + \frac{\partial v}{\partial y}\right), \tag{4.27}$$

where d_h/dt is the horizontal advective derivative.

If the atmosphere is incompressible and barotropic, the continuity equation has:

$$\frac{\partial u}{\partial x} + \frac{\partial v}{\partial y} = -\frac{\partial w}{\partial z},$$

and Equation 4.27, becomes

$$\frac{d_h}{dt}(\zeta + f) = f\frac{\partial w}{\partial z}.$$

If the flow is purely horizontal, $w = 0$ and $\partial w/\partial z = 0$, this reduces to the barotropic vorticity equation

$$\frac{d_h}{dt}(\zeta + f) = 0, \tag{4.28}$$

which returns us to the result we attained in Section 4.5.1.

In frontal zones where the horizontal scale can be as small as 100 km, the approximation leading to Equation 4.27 breaks down because tilting and solenoidal terms become important. The analysis of processes at fronts and other smaller scale phenomena requires different approximations to the equations, but still makes use of the tools we have developed here. Notice that the tools have both diagnostic and analytical uses. They provide useful approaches for analysing meteorological data, as well as being the beginning point of modelling exercises. Ultimately, all work, both research and operational (forecasting), involving large scale atmospheric dynamics is carried out through the implementation and analysis of numerical models of the atmosphere. These models are exemplified by the *numerical weather prediction models* mentioned in Chapter 1.

5

Waves in the atmosphere

It is common to observe patterns in thin, semi-continuous layers of cirrus clouds that are highly suggestive of wave-like motion in the middle and upper troposphere. Astute observers may even have seen clouds that appear to mark a sequence of breaking waves near a large mountain or mountain range. These are indeed indications that waves can exist in the atmosphere, but are a misleading indication because these waves are only evident because they are made visible by clouds whose presence may not be directly related to the waves. There is the possibility that waves could exist in clear air, and also that waves could exist on such a large scale that they are not evident to observers looking up from Earth's surface. An example of a wave-filled atmosphere is shown in Figure 5.1. Studies of atmospheric waves have uncovered a wide range of wave types, all having different dynamical bases, different characteristics, and existing under different atmospheric conditions. All waves have a simple dynamical basis in common – they are driven by restoring forces that act in opposition to a displacement from an equilibrium position. The elasticity of air gives rise to *sound waves*. If the restoring force is gravity, the atmosphere will support *gravity waves*. If the restoring force is both gravity and the Coriolis force, the atmosphere will support *inertia-gravity waves*, while the Coriolis force alone gives rise to *inertial waves*. If the variation of Coriolis force with latitude provides the restoring force, *Rossby waves* will result. These waves are all particular solutions of the governing equations. We will explore only two simple wave types in order to illustrate the approaches needed for their analysis.

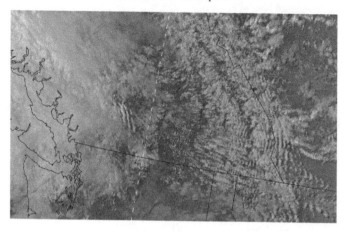

Figure 5.1 A visible band satellite image of a widespread wave field in the lower troposphere over Southern British Columbia. Distinct wave crests a few hundred kilometres long are visible as narrow cloud bands in the centre right of the image. The relatively cloud-free band running NNE–SSW in the midst of the main wave field is due to subsidence over a large valley-like structure called the Rocky Mountain Trench. Image courtesy of NOAA and the University of Washington.

Sound waves are unlike other atmospheric waves in that they are longitudinal waves, in which the oscillation is in the direction of propagation. By contrast, transverse waves have an oscillation perpendicular to their direction of propagation. We will explore internal gravity waves and Rossby waves, but will not deal with sound waves here, other than noting that the speed of sound[1] waves is given by:

$$c_s = \sqrt{\frac{\bar{p}\gamma}{\rho_0}}$$

where γ is C_p/C_v.

5.1 The analysis of propagating waves (briefly)

A simple propagating wave in two dimensions (here we choose the x–z plane) is represented mathematically by:

$$\psi(x,t) = \psi_0 \exp\left[i(kx + mz - \omega t)\right], \tag{5.1}$$

[1] $c_s \approx 340$ m s^{-1} at sea level pressure and $20\,°$C.

where ψ_0 is the amplitude, k and m are the x and z wave numbers respectively; $k = (2\pi/\lambda_x)$ and $m = (2\pi/\lambda_z)$ where λ_x and λ_z are the x and z wavelengths respectively. The speed of wave propagation in the x and z directions is given by $c_x = \omega/k$ and $c_z = \omega/m$ respectively. It is always a given that the physically meaningful solution is $\Re(\psi)$.

The range of permitted combinations of frequencies and wave numbers is given by the *dispersion relation* $\omega(k, m, \pi_1, \pi_2, \ldots)$, where π_i are parameters of the physical system. Real waves in the atmosphere are never of infinite extent in space and time, as implied by Equation 5.1, but rather propagate as a *wave packet* with finite spatial extent and duration. It can be shown that the speed at which such a packet propagates, v_g, known as the *group velocity* is defined by:

$$v_g \equiv \frac{\partial \omega}{\partial k}.$$

A physical interpretation of the group velocity is that v_g is the speed at which energy or information is conveyed by the wave. Since real waves are made up of many components of pure frequency, it is important to consider the speed at which one of the constituent components propagates. This is known as the *phase velocity* v_p, which is defined by:

$$v_p \equiv \frac{\omega}{k}.$$

A physical interpretation of the phase velocity is that v_p is the speed at which a feature (say crest or trough) of a particular wave component travels through space. These ideas will be useful in understanding the direction of propagation of lee waves and Rossby waves treated later in this chapter.

5.2 Simple wave types

The governing equations are strongly non-linear, and admit many complicated wave-like solutions that are inaccessible to linear mathematical analysis. In order to isolate simple wave solutions, we must linearize the equations in a process known as the *perturbation method*. This method starts from the same decomposition of all field variables into mean and fluctuating parts that was introduced in Chapter 3. It is assumed that the mean fields all satisfy the governing equations, and that perturbations themselves are small enough that we may ignore products of perturbations. The non-linear differential equations then become linear differential equations in the perturbation variables. If the perturbations are assumed to be sinusoidal, then standard methods will reveal the properties of linear waves, including their dispersion relation.

5.2.1 Internal gravity waves: speed controlled by stability

As was presented in Chapter 1, in stable atmospheres (or atmospheric layers), buoyancy provides a force that acts in the opposite direction to vertical displacement of an air parcel from its level of equilibrium. Clearly this will be the basis for horizontally propagating transverse waves. Since the restoring force is buoyancy, such waves would naturally be called buoyancy waves, but are by tradition called gravity waves. The adjective 'internal' implies they exist in the interior of a fluid, rather than on a density interface such as waves on the ocean surface. A common example of such waves are *Kelvin–Helmholtz billows* as depicted in Figure 5.2. Anticipating that these waves have short wavelength, we ignore Coriolis forces, and use the governing equations in x and z only. As we only consider small fluctuations, we will neglect the effects of friction. The governing equations, before taking the Boussinesq approximation become:

$$\frac{\partial u}{\partial t} + u\frac{\partial u}{\partial x} + w\frac{\partial u}{\partial z} + \frac{1}{\rho}\frac{\partial p}{\partial x} = 0 \tag{5.2a}$$

$$\frac{\partial w}{\partial t} + u\frac{\partial w}{\partial x} + w\frac{\partial w}{\partial z} + \frac{1}{\rho}\frac{\partial p}{\partial z} + g = 0 \tag{5.2b}$$

$$\frac{\partial u}{\partial x} + \frac{\partial w}{\partial z} = 0 \tag{5.2c}$$

$$\frac{\partial \theta}{\partial t} + u\frac{\partial \theta}{\partial x} + w\frac{\partial \theta}{\partial z} = 0. \tag{5.2d}$$

We now replace variables by the sum of base state and perturbation values, as was done in Section 1.4, but this time involving both horizontal and vertical momentum equations.

$$\begin{aligned}
\rho &= \rho_0(z) + \rho' \\
p &= \bar{p}(z) + p' \\
\theta &= \theta(z) + \theta' \\
u &= u' \\
w &= w'.
\end{aligned} \tag{5.3}$$

Figure 5.2 A sequence of Kelvin–Helmholtz billows, extending over about 20 km and propagating from left to right. Image courtesy of Francis Jones.

We have assumed, without loss of generality, that there is no mean wind in a hydrostatic base state.

Substitution of Equation 5.3 into Equations 5.2a to 5.2d and dropping all terms that contain products of perturbations gives the linearized equations. This will include the Boussinesq approximation in which the last two terms in Equation 5.2b are approximated as:

$$\frac{1}{\rho}\frac{\partial p}{\partial z} + g \simeq \frac{1}{\rho_0}\frac{\partial p'}{\partial z} + \frac{\rho'}{\rho_0}g. \tag{5.4}$$

The potential temperature θ in Equation 5.2d is given by the Poisson equation which can be written as:

$$\theta = \frac{p}{\rho \mathcal{R}} \left(\frac{p_s}{p}\right)^{\mathcal{R}/C_p}. \tag{5.5}$$

Taking the logarithm of the Poisson equation gives:

$$\ln\theta = \frac{1}{\gamma}\ln p - \ln\rho + \text{constant}, \tag{5.6}$$

which in perturbation form becomes

$$\ln\left[\bar{\theta}\left(1+\frac{\theta'}{\bar{\theta}}\right)\right] = \frac{1}{\gamma}\ln\left[\bar{p}\left(1+\frac{p'}{\bar{p}}\right)\right] - \ln\left[\rho_0\left(1+\frac{\rho'}{\rho_0}\right)\right] + \text{constant.}$$

We can now use Equation 5.6 to eliminate the base state pressure, and can use the approximation $\ln(1+\varepsilon) \approx \varepsilon$ if $\varepsilon \ll 1$ to simplify the θ perturbation equation (solved for ρ') as:

$$\rho' \simeq -\rho_0\frac{\theta'}{\bar{\theta}} + \frac{p'}{c_s^2},$$

where $c_s^2 = \bar{p}\gamma/\rho_0$ is the square of the speed of sound. In the buoyancy wave motions we are interested in, the density effect of temperature fluctuations is much greater than the density effect of pressure fluctuations, so:

$$\left|\rho_0\frac{\theta'}{\bar{\theta}}\right| \gg \left|\frac{p'}{c_s^2}\right|,$$

and we may approximate potential temperature fluctuations by:

$$\frac{\theta'}{\bar{\theta}} \simeq -\frac{\rho'}{\rho_0}. \tag{5.7}$$

We can now use Equations 5.4 and 5.7 to develop a linearized set of equations analogous to Equations 5.2a to 5.2d, but for perturbation quantities:

$$\begin{aligned}
\frac{\partial u'}{\partial t} + \frac{1}{\rho_0}\frac{\partial p'}{\partial x} &= 0 \\
\frac{\partial w'}{\partial t} + \frac{1}{\rho_0}\frac{\partial p'}{\partial z} - \frac{\theta'}{\bar{\theta}}g &= 0 \\
\frac{\partial u'}{\partial x} + \frac{\partial w'}{\partial z} &= 0 \\
\frac{\partial \theta'}{\partial t} + w'\frac{\partial \bar{\theta}}{\partial z} &= 0.
\end{aligned} \tag{5.8}$$

With some manipulation, p', u' and θ' can be eliminated from Equation 5.8 to give a single equation for w':

$$\frac{\partial^2}{\partial t^2}\left(\frac{\partial^2 w'}{\partial x^2} + \frac{\partial^2 w'}{\partial z^2}\right) + N^2\frac{\partial^2 w'}{\partial x^2} = 0, \tag{5.9}$$

where $N^2 = (g/\bar{\theta})(\partial\bar{\theta}/\partial z)$ is the Brunt–Väisälä frequency introduced in Chapter 1.

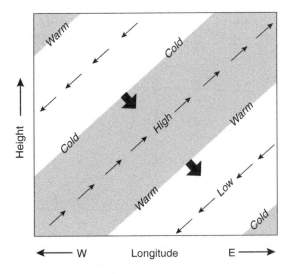

Figure 5.3 Schematic depiction of pressure, temperature and velocity perturbation fields of a pure gravity wave in the x–z plane. Light arrows give the perturbation velocity field, heavy arrows indicate the phase velocity. Here, $k > 0$, $m < 0$ and $\omega > 0$ so that waves tilt upward to the East, and the phase wave propagates downward and to the East. From: Wallace, J. M. and Kousky, V. E., 1968: Observational evidence of Kelvin waves in the tropical stratosphere. *Journal of Atmospheric Science*, **25**, 900–907.

As explained earlier, we now seek solutions by substituting:

$$w'(x,z,t) = \psi_0 \Re \exp[i(kx + mz - \omega t)]$$

into Equation 5.9. This results (after some manipulation) in the dispersion relation:

$$\omega^2(k^2 + m^2) - N^2 k^2 = 0$$

or

$$\omega = \pm \frac{Nk}{\sqrt{k^2 + m^2}}.$$

From this it is clear that the frequency (and hence speed) of the waves varies linearly with N, which is consistent with our earlier analysis of parcel oscillation in the vertical. Figure 5.3 shows schematically the perturbation fields of a propagating gravity wave. Such waves must be triggered by some effect that provides an initial vertical displacement in a stable layer. This is often done by passage of air over a mountain

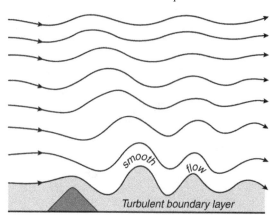

Figure 5.4 Schematic depiction of hypothetical stream lines in a lee wave, triggered by a steep mountain, showing a turbulent boundary layer, three rotors of decreasing size in the lee of the mountain, and vertically propagating gravity waves through the troposphere. These standing waves are tied to the topography, and have crests that tilt upwind at increasing altitude

or mountain range which results in gravity waves in the lee (downwind) of the mountain, hence the common term *lee waves*. Figure 5.4 illustrates hypothetical stream lines in a lee wave. Notice how the wave crests tilt upwind at increasing altitude. Such waves are often observed to propagate vertically through the entire troposphere. If there are layers of moisture at the mountain top elevation, they will condense to form clouds with spectacularly rounded upper surfaces in the wave crests. Because of their lens-like appearance, these are called lenticular clouds, as depicted in Figure 5.5.

5.2.2 Rossby waves: speed controlled by latitudinal variation of the Coriolis force

The upper air (at elevations of 4 to 6 km) flow in mid latitudes of both hemispheres is dominated by a strong current of westerly wind that meanders slightly in the latitude band $40°$ to $60°$. This meandering forms a wave, called the *Rossby wave*, with wavelengths so long that two to five waves fit into a hemisphere. Rossby waves are governed by the magnitude of the mean flow and the rate of change of the Coriolis force with latitude. When the Rossby wave amplitude becomes very

Figure 5.5 Vertically stacked lenticular clouds marking a gravity wave initiated by Mount Sanford, Alaska, USA. Individual lenticular clouds are caused by moisture layering in the atmosphere, the smooth tops marking flow trajectories. Image: Pekka Parviainen/Science Photo Library.

large, masses of cold or warm air detach to become the cyclonic storms that are responsible for day-to-day weather in the mid latitudes.

We now seek two-dimensional (in the x–y plane) wave-like solutions of the linearized barotropic vorticity equation, starting with Equation 4.28, slightly rearranged:

$$\left(\frac{\partial}{\partial t} + u\frac{\partial}{\partial x} + v\frac{\partial}{\partial y} \right) \zeta + v\frac{\partial f}{\partial y} = 0.$$

We linearize around a base state as follows:

$$u = \bar{u}+u', \quad v=v', \quad \zeta = \zeta'.$$

We note that the only mean flow is in the x direction. Since the perturbation field in barotropic atmospheres is divergence free, we can define a perturbation streamfunction ψ' by:

$$u' = -\frac{\partial \psi'}{\partial y}, \quad v' = \frac{\partial \psi'}{\partial x}.$$

This results in $\zeta' = \nabla^2 \psi'$, which gives the desired perturbation equation:

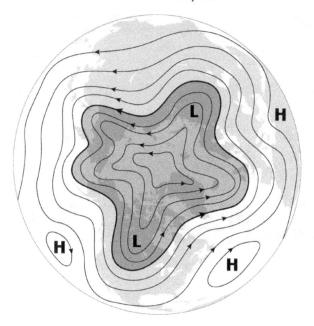

Figure 5.6 Schematic depiction of a sequence of Rossby waves in the Northern Hemisphere. Notice the five lobed structure, and the associated passage of cold polar air/warm subtropical air towards lower/higher latitudes in regions of low/high pressure. Symbols H and L indicate centres of high and low pressure respectively. The contoured variable is the thickness of the 500 HPa level. Arrows indicate the direction of flow at these levels.

$$\left(\frac{\partial}{\partial t} + \bar{u}\frac{\partial}{\partial x}\right)\nabla^2\psi' + \beta\frac{\partial \psi'}{\partial x} = 0, \qquad (5.10)$$

where $\beta \equiv df/dy$ is assumed constant. As before, we seek simple wave solutions by substituting

$$\psi'(x,z,t) = \psi_0 \Re \exp[i(kx + ly - \omega t)],$$

into Equation 5.10 (where k and l are wave numbers in x (E–W) and y (N–S) directions, respectively) to obtain the dispersion relation:

$$\omega = \bar{u}k - \frac{\beta k}{(k^2 + l^2)}.$$

Because $c_x = \omega/k$ the relative wave speed is:

$$c_x - \bar{u} = -\frac{\beta k}{(k^2 + l^2)}.$$

This means that Rossby waves always propagate westward relative to the mean $\bar{u}$ flow, and the phase speed depends on β and both x and y wave numbers. A simple scaling estimate of these Rossby waves shows that they move westward at the modest pace of about 6 m s^{-1} relative to the mean westerly flow. Because of their large scale, Rossby waves are only observable using either multi-station upper air measurements, or indirectly by inference from satellite observations of cloud patterns. Figure 5.6 illustrates schematically a sequence of Rossby waves in the Northern Hemisphere.

This very brief exploration of only two kinds of very simple atmospheric waves has shown the mechanisms underlying gravity (buoyancy) waves and Rossby waves. The simple nature of the perturbation analysis and relative complexity of the resulting dispersion relation suggests that these two types of waves as encountered in nature can be extremely complex. Waves of all types can result in substantial difficulties in numerical prediction because of wave solutions dominating over non-wave phenomena. This problem is dealt with by appropriate filtering of the discretized equations. Probably most remarkable of all is that such complex phenomena can have their essential behaviour revealed by relatively simple mathematical techniques.

Epilogue

This introduction to atmospheric modelling covered a wide range of topics in a sadly short space. In spite of the brevity, I hope to have conveyed some of the most important ideas and approaches employed in mathematical analysis of atmospheric dynamics. In my opinion, the salient ideas are scale based analysis built on an underlying set of equations whose intractability and breadth conspire to prevent simple analysis. That progress can be made using relatively simple mathematics is remarkable. That progress is made is a tribute to the rigour and persistence of a cohort of extraordinary scientists, only some of whom have been named in this book. Ultimately, the subject has an enormously practical expression, that is the societal need for meteorological forecasts. The numerical methods that lie behind modern weather forecasts have as essential foundations the equations I have presented, and the methods I have sketched.

The compact nature of this book, and the specific focus on mathematical modelling have had a less than desirable consequence. The focus on dynamics means that I have not been able to deal with many fascinating and awe inspiring phenomena encountered in the atmosphere. I have not touched on any atmospheric optical phenomena, so I have deprived readers of the wonders and beauty of rainbows, arcs and haloes. I have dealt with neither clouds in all their variety, nor precipitation in its many forms. This means readers will have to look elsewhere to discover the mystery and beauty of a cumulonimbus or a funnel cloud, and will have to broaden their reading to understand the distinction between graupel and hail. I could not ignore atmospheric thermodynamics, but had to give it only enough attention to help in the understanding of atmospheric dynamics. Similarly, radiation in the atmosphere was treated only to the extent that it served an understanding of dynamics. I hope readers will be inspired by what I have presented to undertake their own reading of these omitted topics, and many others.

Probably the most important lesson to be gleaned from this brief tour through various atmospheric phenomena is that simple yet sophisticated models can lay open the essential characteristics of even hugely complex phenomena. As an example, I point to the startling simplicity of the approximate equations that were used to model the Ekman layer. Their simplicity is evident. Their sophistication lies in the fact that they capture the phenomenon and much of its underlying physics. All will acknowledge that the Ekman spiral does not exist in detail in real atmospheres, but nobody will deny that we understand the essential mechanisms underlying many features of the adiabatic, barotropic boundary layer.

I hope by this brief exploration of a selected set of examples to have helped students see the vastly interesting set of phenomena that populate the atmosphere. I hope also to have inspired some of them to undertake more thorough study of these phenomena. There exist many works beyond those presented here and, more important, there remain many fundamental and fascinating problems still to be dealt with. Among those are difficult questions of interactions between scales of motion which I have treated as separate. I am convinced there exist many phenomena that will yield to the application of more sophisticated mathematical approaches than I have used in this introduction. I urge all my readers to develop their mathematical toolbox in as wide a range of techniques as they can.

As I have explained at many points throughout this text, numerical modelling forms a logical and essential next step in atmospheric modelling. Modern atmospheric science is deeply imbedded in numerical modelling. It is my conviction that an intimate interplay between mathematical modelling, observation and numerical modelling forms the essence of a 'three legged stool' upon which we sit to gaze in wonder at the atmosphere. A sad fate will befall gazers who do not maintain all the legs in good working order!

Appendix A

Dimensional analysis and scales

Buckingham pi theorem All measurements are based on one of a limited number of sets of units, called Fundamental Dimensional Units (FDU). Any analytical understanding of relationships between such measurable quantities is constrained by relationships between the various FDUs needed in the measurements. This constraint is best expressed as a set of axioms, or assumptions (depending on your point of view).

(1) An unknown quantity u is dependent on measurable quantities $W_1, W_2, \ldots, W_n$:

$$u = f(W_1, W_2, \ldots, W_n), \tag{A.1}$$

where f is an unknown but desired function of $W_1, W_2, \ldots, W_n$. The central problem in much of science is the discovery of the form of f. This can be done by theorizing, or measurement, often both.

(2) The quantities $W_1, W_2, \ldots, W_n$ are measured in terms of m FDUs which we label as $L_1, L_2, \ldots, L_m$. The choice of FDUs needed to express the sizes of the W and L quantities is up to the researcher. In atmospheric science, the FDUs are: $L_1 = M = $ mass, $L_2 = L = $ length, $L_3 = T = $ time, $L_4 = \theta = $ temperature.

(3) If Z represents any of $(u, W_1, W_2, \ldots, W_n)$, the dimension of Z (written as $[Z]$) is a product of powers of the FDUs:

$$[Z] = L_1^{\alpha_1} L_2^{\alpha_2} \ldots L_m^{\alpha_m}, \tag{A.2}$$

where $(\alpha_1, \alpha_2, \ldots, \alpha_m)$ are real (often restricted to rational) numbers. Z is dimensionless if and only if $[Z] = 1$, all its dimension components are equal to 0.

(4) FDUs are independent of the system of units chosen, but the choice of units can (usually will) result in a scaling of each FDU, but not dimensionless quantities. A fundamental consequence is that Equation A.2 remains unchanged under a change of units.

These apparently simple and seemingly obvious assumptions have far-reaching consequences. Most importantly, if an equation (such as Equation A.1) is to be physically meaningful, it must be dimensionally homogeneous.

Equation A.1 can always be recast in the form

$$\pi = F\left(\pi_1, \pi_2, \ldots, \pi_k\right), \tag{A.3}$$

where all π terms are dimensionless, and k is *reduced* to $k = n - r$.

Equation A.1 has now been replaced by Equation A.3, which is simpler because it has r fewer independent variables. Note that the dimensionless π quantities are given by:

$$\pi = u W_1^{\gamma_1} W_2^{\gamma_2} \ldots W_n^{\gamma_n}$$
$$\pi_i = u W_1^{\gamma_{1i}} W_2^{\gamma_{2i}} \ldots W_n^{\gamma_{ni}}, \quad i = 1, 2, \ldots, k.$$

Simple linear algebra is all that is needed to find the exponents in these two equations. These ideas are called the Buckingham pi theorem.

The reason for the apparent simplification $(k = n - r)$ is that the units of measurement we employ are arbitrary units, assigned to make measurement a simple matter because of a universally accepted scale of measurement. However, each unit of measurement that is introduced increases the dimensionality of the problem (roughly equivalent to the number of variables involved) by one. Problems set in dimensionless form are at their simplest.

Note that the quantity $W_1^{\gamma_1} W_1^{\gamma_1} \ldots W_n^{\gamma_n}$ which is required to render u dimensionless can be thought of as a natural unit of measurement for u, but only in the problem captured in Equation A.1. It is a scale measurement for u that arises naturally out of the problem. To understand this, it is necessary to recognize that all measurement involves division of the measured quantity by the size of the unit of measurement. To say that someone is 1.88 m tall means that if you divide their (unitless) height by the (unitless) length of one metre, the answer will be 1.88. By convention, we append m to the 1.88 to indicate the (arbitrary) unit of measurement used.

Clearly if one is able to approach an atmospheric phenomenon using dimensional analysis, the scales (not just space and time as in Figure 1.1) will arise naturally out of the analysis.

Finally, it is necessary to point out that the Buckingham pi theorem cannot give the functional form of F introduced in Equation A.3. F is usually found by empirical curve fitting, using experimental data.

These ideas may be illustrated by an elementary physical example, that of a simple pendulum. Consideration of the important quantities would lead even an inexperienced physicist to conclude that the period τ of a pendulum depends on its length l, the mass of the pendulum bob m, the acceleration of gravity g and the angle through which it swings θ, so that:

$$\tau = f(m, l, g, \theta). \qquad (A.4)$$

Here $L_1 = M$ (mass), $L_2 = L$ (length) and $L_3 = T$ (time). If we set $W_1 = m$, $W_2 = l$, $W_3 = g$ and $W_4 = \theta$, then $[\tau] = T$, $[m] = M$, $[l] = L$, $[g] = LT^{-2}$, $[\theta] = 1$.

From before, $k = 4 - 3 = 1$ and the problem consists of one dimensionless independent variable, rather than the three dimensional independent variables of Equation A.4. Since the problem as originally stated already has one dimensionless independent variable θ, we may as well take that as the required dimensionless quantity. We must now find the dimensionless dependent variable π such that:

$$\pi = \tau m^{\gamma_1} l^{\gamma_2} g^{\gamma_3} F(\theta).$$

The solution is $\gamma_1 = 0$, $\gamma_2 = -\frac{1}{2}$, $\gamma_3 = \frac{1}{2}$, meaning that τ is independent of m. The period of our simple pendulum is thus:

$$\tau = \sqrt{\frac{l}{g}} F(\theta).$$

The function F can only be determined by experiment. However, a simple experiment involving a single value of $\sqrt{l/g}$ with varying swing amplitude θ will show that $F(\theta)$ is constant with a value of 6.28 $\pm$ (experimental error) for small swing amplitudes. This is the well-known pendulum period $\tau = 2\pi\sqrt{l/g}$.

An alternative view of the pendulum problem is to write

$$\tau_* = F(\theta) = \frac{\tau}{\sqrt{l/g}}$$

as the dimensionless period, if time is measured using the natural unit of time for pendulums, $\sqrt{l/g}$. This unit of time is to be used for all pendulums of length l, on all planets with gravitational acceleration g. Of course in this view, all pendulums have the same dimensionless period, 2π, which is a wonderful simplification that surely would have delighted Galileo!

Bibliography

Cushman-Roisin, B., 1994: *Introduction to Geophysical Fluid Dynamics.* Prentice Hall, NJ, 320 p.

Cushman-Roisin, B. and Beckers, J.-M., 2011: *Introduction to Geophysical Fluid Dynamics: Physical and Numerical Aspects.* Academic Press, Amsterdam, 828 p.

The second of these books is a significantly enlarged and updated revision of the first. A notable enlargement being the addition of numerical exercises to the already excellent analytical problems. Both define geophysical fluid dynamics and outline its importance, emphasizing the role of rotation, waves and stratification. They discuss scales of motion (length and time) and the importance of order-of-magnitude reasoning. They contrast atmosphere and oceans, and discuss the impossibility of formal experiments in both fluid realms. They stress the importance and difficulty of measurements.

Dutton, J., 1976: *The Ceaseless Wind.* McGraw-Hill, New York, 579p.

This is a wonderfully written book by a very wise author. Dutton contrasts heuristic knowledge (understanding) with forecasting (predictions), and emphasizes scale dependence. He discusses the importance of mathematical reasoning. He discusses dynamics driven by thermodynamic forcing, and shows the importance of approximations (hydrostatic, isentropic, geostrophic etc.) to the governing equations. He summarizes the *Grand Problems* as interaction between scales, sensitivity of solutions, climate and modes of response.

Eskinazi, S., 1975: *Fluid Mechanics and Thermodynamics of our Environment.* Academic Press, New York, 422 p.

This book takes a rather different approach than others in this list since it is very much the view of a physicist, rather than that of an atmospheric scientist. The different perspective is refreshing, and makes the book nicely complementary to others in this list. In many places, the author takes particular pains to treat rigorously matters which most atmospheric scientists are content to view in rather pragmatic ways. The result is that the book contains insights that are always useful.

Holton, J. R., 1979: *An Introduction to Dynamic Meteorology*, Second Edition. Academic Press, New York, 391 p.

This book is a classic, and contains the views and perspectives of a master of the subject. It is outstanding for its rigour and breadth, and crystal-clear explanations of a complex subject. It does concentrate mainly on the large scale atmosphere.

Worster, G., 2010: *Understanding Fluid Flow*. Cambridge University Press, Cambridge, 104 p.

This is an excellent, compact introduction to fluid mechanics in general. All the major ideas are introduced and developed in an accessible and logical way. As it is in the Cambridge University Press AIMS Library Series, I assume the intended audience for the present book will have easy access to it. I recommend all readers of this book to first study Worster.

Index